Contents

	Introduction	1
1	Today's Mathematics Classrooms	5
2	Productive Partnerships	23
3	Using Hands-on Moveable Objects	35
4	Interacting with Web-Based Resources	71
5	Playing Games and Solving Puzzles	99
	Conclusion	133
	About the Author	139

Introduction

Well-planned and well-implemented parental involvement initiatives contribute to students' success in mathematics (Haycock, 2001), teachers' increased comfort levels collaborating with parents (Epstein & Sanders, 1998), and better home-school relations (Epstein, Salinas, & Jackson, 1995). Despite these findings, fear of losing control of classroom teaching, school management, and professional status have surfaced in all countries as the main reasons for some teachers avoiding parental involvement (Epstein & Sanders, 1998).

In such situations, there is an absence of guidelines and materials for parent workshops and initiatives that can in turn cause parents' lack of engagement in their children's learning of mathematics. My intention in writing this book is to enliven three-way partnerships among parents, teachers, and students concerning mathematical learning.

I present initiatives that nurture productive parent-child collaboration by building parents' understanding of the best practices in the mathematics classroom, facilitating mathematics learning at home, and maintaining connections between the mathematics classroom and the home.

Parental involvement in mathematics education is certainly not new for teachers, but it is a component that must remain energized since studies have shown that it is critical for teachers to partner with parents to achieve the goal of students succeeding in mathematics (Bezuk, Whitehurst-Payne, & Aydelotte, 2000). Key tenets of the principles concerning constructivism and overlapping spheres of influence are presented in the first two chapters to provide a solid theoretical basis for teaching mathematics the way we do and for involving parents in the learning process.

A review of research is presented in chapter 1 concerning changes in mathematics teaching over the years, the characteristics of present-day constructivist mathematics classrooms, the need for parental involvement, and parental issues and perspectives that warrant attention to productively involve parents in mathematics education reform efforts. Key points are included at the end of chapter 1 to serve both as a review of the material discussed and as organized information for teachers to share with parents.

Chapter 2 includes a discussion about the theory of overlapping spheres of influence that, along with relevant research findings, helps provide the rationale for involving parents in mathematics education and supports the model framework from which this book's parental involvement initiatives stem. Effective ways to partner with parents and the features that permeate the initiatives discussed in later chapters are explained. As in chapter 1, key points are included for teachers to share with parents.

Rather than generalizing out of context, this book integrates in its remaining chapters much of its theory and practical advice to readers by way of easy-to-use activities and guidelines for parent mathematics initiatives that I used successfully in various school communities. It is my hope that teachers will use them as well as design their own using the model to excite both parents and their children about doing mathematics together.

The initiatives offer parents and children opportunities to learn from each other through engagement in mathematics activities that build both parents' awareness of content and best practices embodied in the *Principles and Standards for School Mathematics* (NCTM, 2000) and children's awareness of their parents as partners in their learning of mathematics.

Each initiative reflects a model consisting of an invitation, an initial meeting, an engagement workshop, home activities, a follow-up session, and additional tasks to maintain connections between the mathematics classroom and the home.

Integrated into the model are features that (1) *inform* the parent community about the importance of their involvement and the rationale behind specific methods used in the mathematics classroom; (2) *engage* both parents and children in mathematics learning activities at workshop settings and in the home; (3) promote *reflection* as a community about their learning experiences; and (4) *maintain connections* between the home and the mathematics classroom through use of interactive homework, classroom follow-up techniques, and interactive newsletters.

It is emphasized that teachers share material and activities with parents that is relevant not only to their child's grade level but also for all grade levels (Pre-K to 8) to promote a global sense of mathematics learning across the grades. Specific topics of the initiatives are hands-on moveable objects, web-based resources, games, and puzzles used in the learning of mathematics. These selected topics are not meant to be interpreted as an exhaustive list of ways to involve parents but rather as a sampling of my own field-tested action steps that send the message of reform-based mathematics education to parents.

This resource is meant for use by teachers in the field and educators in teacher preparation programs, administrator programs, and in other courses and workshops that prepare professionals to work in mathematics classrooms and with families.

It can be used in conjunction with a methods text or as a supplementary text in courses on mathematics education at elementary and middle school levels. Chapters may also be selected for courses in sociology of education, practicum teaching seminars, educational administration, community education, and staff development workshops.

REFERENCES

Bezuk, N. S., Whitehurst-Payne, S., & Aydelotte, J. (2000). Successful collaborations with parents to promote equity in mathematics. In W. G. Secada (Ed.), *Changing the faces of mathematics*. Reston, VA: NCTM, 143–48.

Epstein, J. L., Salinas, K., & Jackson, V. (1995). *Tips (Teachers involve parents in schoolwork)*. Baltimore, MD: Center on School, Family, and Community Partnerships, Johns Hopkins University.

Epstein, J. L., & Sanders, M. (1998). International perspectives on school-family community partnerships. *Childhood Education*, 74(6), 340–41.

Haycock, K. (2001). Closing the achievement gap. *Educational Leadership*, 58(6), 6–11.

National Council of Teachers of Mathematics (NCTM). (2000). *Principles and standards for school mathematics*. Reston, VA: Author.

Chapter One

Today's Mathematics Classrooms

Learning at home involves parents monitoring and assisting their children with tasks including homework and other curricular-linked activities and decisions. However, with changing mathematics content and pedagogy, parental assistance at home may unintentionally undermine what the mathematics teacher is trying to accomplish in the classroom (Peressini, 1997).

During my years teaching mathematics, I as well as my colleagues have often encountered parents questioning the way the subject was taught because they found the methods and content different from what they experienced as learners of mathematics.

They worried whether their children were learning properly and if they would be able to assist them at home. This sense of unfamiliarity often resulted in either a lack of home involvement with mathematics learning or input that contradicted teachers' classroom practices.

Eighteen years ago, when I began teaching undergraduate and graduate mathematics methods courses, I started integrating strategies for informing and engaging parents in current trends of teaching mathematics. This component of my courses grew extensively as a result of my experiences traveling to various schools and discussing with parents their issues and perspectives concerning mathematics education.

Their viewpoints are shared in this chapter along with research that both supports parental involvement in mathematics education and illuminates areas that warrant attention to assure that parental involvement is productive. The information presented in each part of this chapter is summarized in a section titled "Key Points to Share with Parents" in a manner that lends itself for use at parent events. References to parents are meant to include all adults who play an important caretaker role in a child's home life.

MATHEMATICS TEACHING THROUGH THE YEARS

Mathematics teaching in the United States has changed dramatically throughout the years (Usiskin & Dossey, 2004). To gain parental support for such changes, it is important to build parents' knowledge of where mathematics teaching has been and where it is now. Otherwise, the quality of today's teaching practices cannot be fully appreciated by parents.

A description of the changes that have taken place in mathematics teaching is presented in this section to provide a historical record to share with parents about how current mathematics classrooms have evolved. The described changes stem from workshops given by Cathy Seeley, a former president of the National Council of Teachers of Mathematics (NCTM).

Many say that knowledge is power. It is with this thought in mind that the following information is shared as we seek to empower parents as partners in their child's learning of mathematics. We'll start in the 1940s, a period that has been described as a time of complacency where drill and practice methods of teaching permeated classrooms and were accepted without debate.

Concern arose though in the 1950s, known as a period of awakening, when the Russian satellite Sputnik was launched. This event in history caused Americans to see themselves as taking a backseat to the accomplishments of others, and left educators questioning their existing mathematics curriculum and pedagogical practices.

Consequently the 1960s, sometimes referred to as a time of overreaction, were filled with changes in teaching styles. Due to educators disturbed about the thought of their students viewing mathematics as a subject consisting only of isolated bits of information that needed to be memorized, drill and practice exercises were replaced with teaching methods that nurtured conceptual understanding.

Unfortunately, computational accuracy became underemphasized causing this state of overreaction to continue into the 1970s. Educators became concerned over students' lack of efficiency with computational tasks, as well as their lack of meaningful understanding of computational algorithms and other mathematical ideas.

Rote memorization of mathematics facts once again gained a presence, leaving the 1980s to be viewed as a period of turmoil. Two schools of thought, consisting of drill and practice versus conceptual understanding, debated over best practices. The resolution came in 1989 with the NCTM shedding light on the fact that, for students to learn mathematics in a meaningful manner, they need *both* conceptual understanding and computational accuracy (NCTM, 1989).

NCTM energized curriculum and evaluation standards for teachers of grades Pre-K to 12 by providing the framework for content, lesson structure, and assessment techniques that set goals and sought change. Reform efforts moved instruction away from a transmission model where passive learning and a telling pedagogy exist, to a constructivist model where the learner actively participates with the teacher to solve problems, engage in inquiry, and construct knowledge (Grant, 1998; Noddings, 1993).

Debate concerning best practices emerged in the 1990s. Issues such as small group versus whole group instruction, authentic versus standardized assessment, and calculator use versus paper and pencil computation were discussed. As in 1989, NCTM found that *both* sides of each issue had their place in the mathematics arena. With the perspective in mind that reform is an ongoing process that needs to be periodically examined, evaluated, field-tested, and revised, NCTM crafted the *Principles and Standards for School Mathematics* (NCTM, 2000).

This document consisting of curriculum and teaching objectives marked the beginning of focus and commitment to mathematics education. It enhanced what was established in 1989 and energized the movement toward constructivist teaching with the basic message that all students must learn meaningful mathematics using appropriate tools.

NCTM calls upon all of today's mathematics educators to provide a complete curriculum to every student that relates mathematics to our world and builds conceptual understanding, computational fluency, and problem-solving ability with a variety of concrete moveable objects (manipulatives).

The recommended mathematics topics described in *Principles and Standards for School Mathematics* are Number and Operations, Algebra, Geometry, Measurement, Data Analysis and Probability, Problem Solving,

Reasoning and Proof, Communication, Connections, and Representation. I recommend that you review the actual document created by NCTM to enhance your discussions with parents. The document is cited in the bibliography for this chapter and can also be found at www.nctm.org.

NCTM understands that with the responsibility of delivering such meaningful mathematics experiences comes the challenge of planning instructional time so that important mathematical topics are taught effectively and in depth. In response to such a challenge, NCTM recently presented *Curriculum Focal Points for Prekindergarten through Grade 8 Mathematics: A Quest for Coherence* (NCTM, 2006).

This document extends the leadership provided by NCTM over the years by providing areas of emphasis within each grade from prekindergarten through grade 8. Such an approach provides clear, consistent priorities and focus that serve to provide direction for a coherent mathematics curriculum across the United States.

The key focal topics targeted in *Curriculum Focal Points for Prekindergarten through Grade 8 Mathematics: A Quest for Coherence* are Number and Operations, Algebra, Geometry, Measurement, and Data Analysis. It is stressed that these topics be presented in ways that nurture problem solving, reasoning, communication, connections, and representations. I recommend that you review this actual document as well to enhance your discussions with parents. It is also cited in the bibliography for this chapter and can be found at www.nctm.org.

A CONSTRUCTIVIST FOUNDATION TO TEACHING MATHEMATICS

The constructivist theory of learning is the driving force behind NCTM's reform efforts. Basic to constructivism is the idea that knowledge is an individual construction created by the learner as he or she interacts with people and things in the environment (Koch, 2005). Constructivist theorists Jean Piaget (1974), Jerome Bruner (1986), and Zoltan Dienes (1960) agree that abstract thinking of mathematical ideas is possible only after conceptualization and meaningful understanding have been established.

To help build parents' understanding of such a theory in present-day mathematics learning, information to share with them concerning the characteristics of constructivist mathematics classrooms and supporting research for such environments are discussed in this section. Just as our students require reasons for what and how they learn mathematics to give purpose for their efforts, parents require the same consideration to give purpose to their supporting role in the learning process.

Vygotsky's theoretical contributions to the development of curricula and pedagogy indicate that the quality of instruction is crucial in terms of the effect it produces on cognitive development (Moll, 1990). Steele (1999), under the guidance of Vygotsky's sociocultural theory, states that children's level of mathematics development provides a "window of opportunity" for a range of learning activities.

The lower limit of this window rests on previous concepts and skills that have been established at prior levels of learning. The upper limit is determined by the tasks and learning environment that the present teacher facilitates.

This understanding of the relationship between cognitive development and instructional design emphasizes the importance of providing consistent constructivist pedagogical practices throughout the grades so that learning experiences build upon each other in a meaningful manner. Teachers who foster constructivist learning experiences realize that stages of thinking exist and children progress through these stages from concrete manipulations to abstract symbolic representations (Reys et al., 2001).

Teachers using a developmental instructional approach actively involve children in building new knowledge based on experience as opposed to an explanatory approach that tends to promote passive learning and builds dependence on a teacher or textbook (Cathcart et al., 2003). Students are encouraged to take risks and explore mathematics concepts. Their curiosity is tapped into and their thinking stimulated as they engage in doing mathematics to discover patterns and relationships.

Mistakes are viewed as opportunities for learning as teachers guide students to use their incorrect thinking to correct themselves and deepen their understanding. It is also commonplace for children to solve mathematics problems that have more than one right answer and more than one correct method of solution as they broaden their understanding of mathematics by approaching problems from different perspectives.

Communication is a key element in constructivist classrooms as well because written and verbal communication allows students to gather, organize, and clarify thoughts in a manner that contributes significantly to conceptual understanding (Flores & Brittain, 2003).

Instead of filling instructional time with children silently completing workbook pages or taking timed tests on basic facts, teachers who implement constructivist pedagogical practices enliven their classrooms with lively debate about different ways to accomplish mathematics tasks through thinking strategies and concrete explorations. Such engagement supports students' growing ability to strategize and share thinking that is clearly presented and increasingly efficient.

Constructivist learning environments are intrinsically rewarding and self-generative as well. They enhance memory, facilitate the learning of new concepts and procedures, as well as improve problem-solving abilities, attitudes, and beliefs about mathematics (Van de Walle, 2001; Drake, Spillane, & Huffred-Ackles, 2001). Such results legitimize the significant changes in the academic tasks students now engage in when learning mathematics, the ways they interact in classrooms, and what it means to know mathematics.

Instruction has moved beyond "mechanistic implementation" of mathematical procedures and now focuses on building conceptual understanding of these procedures as well (Loucks-Horsley & Matsumoto, 1999; Benken & Brown, 2004). To summarize, today's mathematics classrooms provide inquiry-based instruction where mathematics is viewed as a tool for thought, rather than a set of rules and procedures to be memorized.

THE NEED FOR PARENTAL INVOLVEMENT

For most parents, the reform efforts affecting today's mathematics classrooms have created a learning environment that is very different from what they experienced as learners of mathematics (Mistretta, 2004). This unfamiliar environment for parents often results in confusion and anxiety about how to help their children.

Price (1996) reports on a former president of the National Council of Teachers of Mathematics stating that "we have to help parents bridge their fear and encourage them to join hands in providing a solid mathematics education for all students" (p. 538). Part of building such a bridge involves strengthening parents' awareness of their vital role in mathematics education.

This section highlights research that supports parental involvement in children's learning of mathematics. We as mathematics educators need to share this information with parents so that they can productively collaborate with their children. When parents encourage and support mathematics learning, children have an added advantage in school. Their input has shown to be a factor in achieving the goal of all students succeeding in mathematics (De La Cruz, 1999; Haycock, 2001; Bezuk, Whitehurst-Payne, & Aydelotte, 2000, Peressini, 1998).

Studies find that parental involvement that supports home-school connections in mathematics learning on a regular basis not only provides a strong foundation for children's attitudes toward mathematics (Kliman, 1999) but also serves as a catalyst in promoting mathematics understanding (Kokoski & Downing-Leffler, 1995) and higher academic achievement in mathematics (Goldstein & Campbell, 1991).

Partnering allows the entire learning community to benefit as everyone joins together to make sense out of mathematics. For example, students' mathematics abilities improve when they engage in discussion about their mathematics thinking (Ford & Crew, 1991) and work cooperatively with a parent when problem solving (Epstein & Dauber, 1991; Myers, 1985).

As parents' understanding of the changes occurring in school mathematics increases, they begin to strengthen their school's efforts to reform mathematics programs (Peressini, 1997) through the realization of the opportunities they have for creating mathematics moments with their children (Moldavin, 2000). In turn, teachers benefit from families serving as partners that support and reinforce mathematical learning at home (Epstein & Sanders, 1998).

PARENTAL ISSUES AND PERSPECTIVES

Low student academic achievement has been found to correlate with negative family attitudes and beliefs about mathematics (Klinger, 2000). For this reason, it is important as mathematics educators to be aware of existing challenges and misperceptions so that our actions target areas that warrant attention. This section illuminates such areas for you as you seek to serve and guide parents, and in turn positively affect your students.

Parents who did well in mathematics when it was taught in a drill and practice manner often feel uneasy about changing a learning environment that worked for them. Parents who did not do well in mathematics are also concerned that their weaknesses may transmit to their children (Burns, 1998). Emerging frustration from the belief that only the specially trained can become involved in mathematics education (Bristor, 1987) creates additional anxiety as well.

Mistretta (2004, p. 72) reports parent pleas such as follows: "Math in our house is a nightmare"; "We need help!" "I don't know how to help my child understand concepts with manipulatives"; "Mathematics today is taught differently than in my time; I don't want to confuse my child"; and "I don't want to show them my way; it may be the wrong way."

Burns (1998, p. ix) has discovered what she calls "mathematical myths" permeating throughout society that include the following: "Only some people are good in math"; "You're only good in math if you have the mathematics gene"; "People who are good in math wear thick eyeglasses and plastic pocket protectors"; and "Mathematicians are different from most of the population."

To add to this situation of falsities, a good number of parents, especially mothers, have been found to believe that learning mathematics only entails recalling basic facts and doing lots of computational exercises, rather than the development of student thinking that is the focus of present-day reform efforts (Warren & Young, 2002).

Unsettling perspectives concerning gender exist as well. Although boys and girls do equally well in school, many parents tend to view boys as superior to girls in mathematics (Raty, Vanska, & Karkkainen, 2002). They often credit boys' successes in mathematics to innate talent, while girls' successes are seen as due to the effort they put into the subject. Many parents of young boys tend to expect their sons to develop mathematical skills earlier than parents of young girls (Carraher, Carraher, & Schliemann, 1985).

Parents of older children often believe that their daughters must work harder to achieve good grades in mathematics and stress the importance of mathematics only with their sons (Leedy, LaLonde, & Runk, 2003). Boys and girls in turn develop different skills, knowledge, and motivation due to such interactions with their parents.

Cause for concern in this area is understandable since children are aware of their parents' expectations (Entwisle & Baker, 1983), and this awareness predicts their performance (Carr, Jessup, & Fuller, 1999). When girls and boys receive different messages about themselves as mathematicians, their attitudes and approaches to mathematics are influenced (Parsons, Adler, & Kaczala, 1982).

An example of such an effect was revealed in a study that found girls less confident in their mathematics abilities due to a perception that their mothers have low expectations for their success in mathematics (Leedy, LaLonde, & Runk, 2003).

These ways of thinking do not serve children well in developing the kind of mathematical understanding and skills needed in today's classrooms. Parents must not pass on negative messages to their children. Doing so can convince children that they cannot be successful in mathematics, or can provide an easy excuse for not even trying (Burns, 1998). Rather, parents need to help deliver a more positive message about what math can and should be to all of us.

Parents who praise the drill and practice model that they may have experienced as students must realize that children need to develop efficient thinking strategies for arithmetical competency as well through use of practices such as modeling and real-life applications.

Parents who see themselves as weak in mathematics must understand that ability in mathematics is not hereditary, and their children are more likely to succeed in mathematics if they support and provide opportunities to conceptually understand mathematics.

Ultimately, parents must realize that mathematics is a way of thinking, and that most of us are perfectly capable of learning mathematics with understanding and with pleasure. It all depends on the learning environment in which we are engaged.

Most parents, including full-time working parents, desire to be involved in their children's education and want to learn how to help their children reach important goals (Epstein & Jansorn, 2004). Unfortunately, parents often don't know how to contribute to their children's learning of mathematics (Burns, 1998).

A common request from parents is extra worksheets to be completed after school. This is counterproductive because children need time to relax after a long day at school (Ford & Crew, 1991). Even more importantly, learning at home needs to reflect the developmental hands-on learning of the classroom. Watching and correcting worksheet completion is not productive learning at home. Rather, it can send an incorrect message to children about mathematics and the role of their parents.

To create productive collaboration, parents need to engage in workshops that facilitate understanding about reform efforts in mathematics education (Sheldon & Epstein, 2001). Parents often want to know how to help their children at home (Dauber & Epstein, 1993; Epstein, 1986) and ask for hands-on workshops so they can concretely learn how to use the tools that their children use to learn mathematics (Mistretta, 2004; Orman, 1993).

Research also indicates that parents are much more knowledgeable about their children's mathematical thinking when they engage in mathematical tasks with their child at workshop settings (Tregaskis, 1991) and when homework encourages parent-child collaboration (Cathcart et al., 2003).

The roles that parents play in their children's learning of mathematics clearly need empowerment so that parents can better serve as links between home and school that enhance learning and build confidence. Parents need to see mathematics as a subject they can enjoy with their children that goes beyond just checking homework. Later chapters of this book serve to guide the ways we empower parents through initiatives that involve parents and promote mathematics as a subject that can be engaging, even fun.

KEY POINTS TO SHARE WITH PARENTS

Summarized information is presented at this point to help spark conversations with parents. I suggest using this information to create transparencies, PowerPoint presentations, and/or hand-outs during the parental involvement initiatives discussed in later chapters so that you have a guided script and parents have materials to refer to both during an initiative's events as well as at home.

To avoid repetition in cases where multiple initiatives are conducted with the same group of parents, this information can be shared once and then reviewed on subsequent occasions or adapted according to individual teachers' discretion.

Mathematics Teaching through the Years

- 1940s Complacency (Drill and Practice Methods)
- 1950s Awakening (1957—Sputnik)
- 1960s Overreaction (Conceptual Learning)
- 1970s Overreaction (Back-to-Basics)
- 1980s Resolution (Drill and Practice and Conceptual Learning)
- 1990s Debate (What instructional practices to use?)
- 2000s Focus and Commitment (All students doing meaningful mathematics using appropriate tools)
    - *Principles and Standards for School Mathematics*
    - *Curriculum Focal Points for Prekindergarten through Grade 8 Mathematics: A Quest for Coherence*

A Constructivist Foundation to Teaching Mathematics

- Knowledge is an individual construction created by the learner as he/she interacts with people and things in the environment.
- Stages of thinking exist, and children progress through these stages from concrete manipulations to abstract representations.
- A developmental instructional approach is used where children actively build new knowledge on experience.
- Thinking is stimulated as children engage in discovery of patterns and relationships.
- It is commonplace for children to solve mathematics problems that have more than one correct answer using more than one correct method of solution.
- Children communicate in verbal and written form about their mathematics thinking.

The Need for Parental Involvement

- Schools that have strong parental involvement have students who learn more and perform better.

- The benefits of parental involvement include an increase in student achievement, a stronger parent-child relationship, and an added resource to teachers.
- Parental involvement is an important factor in producing higher academic achievement specifically in mathematics.
- Consistent parental involvement in mathematics strengthens children's conceptual understanding and attitudes.
- Students' mathematics abilities improve when they are able to discuss their thinking with a family member.

Mathematics "Misbeliefs"

- Mathematics only entails recalling basic facts and doing lots of exercises.
- You're only good in math if you have the mathematics gene.
- People who are good in math wear thick eyeglasses and plastic pocket protectors.
- Mathematicians are different from most of the population.
- Boys do better in mathematics than girls.

How Parents Can Help

- View mathematics as a subject that can be engaging, even fun!
- Get involved with your child's learning of the subject.
- Realize children need more than just drill and practice.
- Realize children are more likely to succeed in mathematics when they develop efficient thinking strategies through use of modeling and real-life applications.
- Learning at home needs to reflect the developmental, hands-on learning of the classroom, not solely the completion of a series of worksheets and checking of homework.
  - Create mathematical/teachable moments where your children are engaged while you observe to learn about their thinking.
  - Share your ideas/methods of solution as well as listen to those of your children.
- Don't pass on a negative attitude. Your mathematics ability or lack of it is not hereditary.

- Help deliver a positive message about what mathematics can and should be to all of us.

SUGGESTIONS FOR FURTHER READING

Anderson, A. (1997). Families and mathematics: A study of parent-child interactions. *Journal for Research in Mathematics Education*, 28(4), 484–512.

Beedle, M. (2003). Spark your child's success in math and science: Practical advice for parents. *Teaching Children Mathematics*, 9(2), 430.

Ford, M., Follmer, R., & Litz, K. (1998). School-family partnerships: Parents, children, and teachers benefit! *Teaching Children Mathematics*, 4(6), 310–13.

Kellermeier, J. (2000). Mathematics, gender, and culture. *Transformations*, 11(2), 41.

Mirra, A. (Ed.). (2004). *A family's guide: Fostering your child's success in school mathematics*. Reston, VA: NCTM.

Morgan, S., & Sorenson, A. (1999). Parental networks, social closure, and mathematics learning: A test of Coleman's social capital explanation of school effects. *American Sociological Review*, 64(5), 661–81.

National Council of Teachers of Mathematics (NCTM). (1999). *Changing the faces of mathematics: Perspectives on Asian Americans and Pacific Islanders*. Reston, VA: Author.

National Council of Teachers of Mathematics (NCTM). (1999). *Changing the faces of mathematics: Perspectives on Latinos*. Reston, VA: Author.

National Council of Teachers of Mathematics (NCTM). (2000). *Changing the faces of mathematics: Perspectives on African Americans*. Reston, VA: Author.

National Council of Teachers of Mathematics (NCTM). (2000). *Changing the faces of mathematics: Perspectives on multiculturalism and gender equity*. Reston, VA: Author.

National Council of Teachers of Mathematics (NCTM). (2001). *Changing the faces of mathematics: Perspectives on gender*. Reston, VA: Author.

National Council of Teachers of Mathematics (NCTM). (2002). *Changing the faces of mathematics: Perspectives on indigenous people of North America*. Reston, VA: Author.

Reutter, J. (2002). How to help your child excel in math: An a to z survival guide. *Mathematics Teaching in the Middle School*, 8(2), 123–24.

Schiller, K., Khmelkov, V., & Wang, X. (2002). Economic development and the effects of family characteristics on mathematics achievement. *Journal of Marriage and Family*, 64(3), 730–42.

Scholastic, Inc. (1990). *Scholastic explains math homework: Everything children (& parents) need to survive 2nd and 3rd grades.* Scranton, PA: HarperCollins Publishers.

Wickelgren, W. (2001). *Math coach.* New York: Berkley Publishing Group.

REFERENCES

Benken, B., & Brown, N. (2004). Improving students' mathematical understandings: An exploration into teacher learning in an urban professional development setting. Paper presented at the North American Chapter of the International Group for the Psychology of Mathematics Education, Toronto, Canada.

Bezuk, N. S., Whitehurst-Payne, S., & Aydelotte, J. (2000). Successful collaborations with parents to promote equity in mathematics. In W. G. Secada (Ed.), *Changing the faces of mathematics.* Reston, VA: NCTM, 143–48.

Bristor, V. J. (1987). But I'm not a teacher. *Academic Therapy,* 23(1), 23–27.

Bruner, J. (1986). *Actual minds, possible worlds.* Cambridge, MA: Harvard University Press.

Burns, M. (1998). *Math: Facing an American phobia.* Sausalito, CA: Math Solutions Publications.

Carr, M., Jessup, D., & Fuller, D. (1999). Gender differences in first-grade mathematics strategy use: Parent and teacher contributions. *Journal for Research in Mathematics Education,* 30(1), 20–47.

Carraher, T. N., Carraher, D. W., & Schliemann, A. D. (1985). Mathematics in the streets and in the schools. *British Journal of Developmental Psychology,* 3, 21–29.

Cathcart, W., Pothier, Y., Vance, J., & Bezuk, N. (2003). *Learning mathematics in elementary and middle schools.* Upper Saddle River, NJ: Pearson Education, Inc.

Dauber, S. L., & Epstein, J. L. (1993). Parents' attitudes and practices of involvement in inner-city elementary and middle schools. In N. Chavkin (Ed.), *Families and schools in a pluralistic society.* Albany, NY: SUNY Press, 53–71.

Day, C. (2000). Teachers in the twenty-first century: time to renew the vision. *Teachers and Teaching: Theory and Practice,* 6, 101–15.

De La Cruz, Y. (1999). Reversing the trend: Latino families in real partnerships with schools. *Teaching Children Mathematics,* 5(5), 296–300.

Dienes, Z. (1960). *Building up mathematics.* London: Hutchinson Education.

Drake, C., Spillane, J., & Huffred-Ackles, K. (2001). Storied identities: teacher learning and subject-matter context. *Journal of Curriculum Studies,* 33(1), 1–23.

Entwisle, D. R., & Baker, D. P. (1983). Gender and young children's expectations for performance in arithmetic. *Developmental Psychology*, 19, 200–209.

Epstein, J. L. (1986). Parents' reactions to teacher practices of parent involvement. *Elementary School Journal*, 86(3), 277–93.

Epstein, J. L., & Dauber, S. (1991). School programs and teacher practices of parent involvement in inner-city elementary schools. *Elementary School Journal*, 91(3), 289.

Epstein, J. L., & Jansorn, N. (2004). School, family, and community partnerships link the plan. *Education Digest*, 69(6), 19–23.

Epstein, J. L., & Sanders, M. (1998). International perspectives on school-family community partnerships. *Childhood Education*, 74(6), 340–41.

Flores, A., & Brittain, C. (2003). Writing to reflect in a mathematics methods course. *Teaching Children Mathematics*, 10(2), 112–18.

Ford, M., & Crew, C. (1991). Table-top mathematics: A home-study program for early childhood. *Arithmetic Teacher*, 38(8), 6–12.

Goldstein, S., & Campbell, F. (1991). Parents: A ready resource. *Arithmetic Teacher*, 38(6), 24–27.

Grant, S. G. (1998). *Reforming reading, writing, and mathematics: Teachers' responses and the prospects for systemic reform*. Mahwah, NJ: Erlbaum.

Haycock, K. (2001). Closing the achievement gap. *Educational Leadership*, 58(6), 6–11.

Kliman, M. (1999). Beyond helping with homework: Parents and children doing mathematics at home. *Teaching Children Mathematics*, 6(3), 140–46.

Klinger, D. (2000). Hierarchical linear modeling of student and school effects on academic achievement. *Canadian Journal of Education*, 25(2), 41–55.

Koch, J. (2005). *Science stories*. Boston: Houghton Mifflin Company.

Kokoski, T., & Downing-Leffler, N. (1995). Boosting your science and math programs in early childhood education: Making the home-school connection. *Young Children*, 50(5), 35–39.

Leedy, M., LaLonde, D., & Runk, K. (2003). Gender equity in mathematics: Beliefs of students, parents, and teachers. *School Science and Mathematics*, 103(6), 285–92.

Loucks-Horsley, S., & Matsumoto, C. (1999). Research on professional development for teachers of mathematics and science: The state of the scene. *School Science and Mathematics*, 99(5), 258–71.

Mistretta, R. M. (2004). Parental issues and perspectives concerning mathematics education at elementary and middle school settings. *Action in Teacher Education*, 26(2), 69–76.

Moldavin, C. (2000). A parent's portfolio: Observing the power of Matt, the mathematician. *Teaching Children Mathematics*, 6(6), 372–75.

Moll, L. C. (Ed.). (1990). *Vygotsky and education: Instructional implications and applications of sociocultural psychology.* New York: Cambridge University Press.

Myers, J. (1985). *Involving parents in middle level education.* Columbus, OH: National Middle School Association.

National Council of Teachers of Mathematics (NCTM). (1989). *Curriculum and evaluation standards.* Reston, VA: Author.

National Council of Teachers of Mathematics (NCTM). (2000). *Principles and standards for school mathematics.* Reston, VA: Author.

National Council of Teachers of Mathematics (NCTM). (2006). *Curriculum focal points for prekindergarten through grade 8 mathematics: A quest for coherence.* Reston, VA: Author.

Noddings, N. (1993). Constructivism and caring. In R. D. Davis and C. A. Maher (Eds.), *School mathematics, and the world of reality.* Boston: Allyn & Bacon, 35–50.

Orman, S. (1993). Mathematics backpacks: Making the home-school connection. *Arithmetic Teacher*, 40(6), 306–309.

Parsons, J. E., Adler, T. F., & Kaczala, C. M. (1982). Socialization of achievement attitudes and beliefs: Parental influences. *Child Development*, 53, 310–21.

Peressini, D. (1997). Parental involvement in the reform of mathematics education. *The Mathematics Teacher*, 90(6), 421–27.

Peressini, D. (1998). The portrayal of parents in the school mathematics reform literature: Locating the context for parental involvement. *Journal for Research in Mathematics Education*, 29(5), 555–76.

Piaget, J. (1974). *The child and reality: Problems of genetic psychology.* London: Frederick Muller (original work published 1972).

Price, J. (1996). President's report: Building bridges of mathematical understanding for all children. *Mathematics Teacher*, 89(6), 536–39.

Raty, H., Vanska, J., & Karkkainen, R. (2002). Parents' explanations of their child's performance in mathematics and reading: A replication and extension of Yee and Eccles. *Sex Roles*, 46(3/4), 121–28.

Reys, R., Lindquist, M., Lambdin, D., Smith, N., & Suydam, M. (2001). *Helping children learn mathematics.* New York: John Wiley & Sons, Inc.

Sheldon, S., & Epstein, J. L. (2001). Focus on math achievement: Effects of family and community involvement. Paper presented at the 2001 annual meeting of the American Sociological Association, Anaheim, CA.

Steele, D. (1999). Learning mathematical language in the zone of proximal development. *Teaching Children Mathematics*, 4(1), 38–42.

Tregaskis, O. (1991). Parents and mathematical games. *Arithmetic Teacher*, 38(7), 14–17.

Usiskin, Z., & Dossey, J. (2004). *Mathematics education in the United States 2004*. Reston, VA: NCTM.

Van De Walle, J. (2001). *Elementary and middle school mathematics*, 4th ed. New York: Addison Wesley Longman, Inc.

Warren, E., & Young, J. (2002). Parent and school partnerships in supporting literacy and numeracy. *Asia-Pacific Journal of Teacher Education*, 30(3), 217–28.

Chapter Two

Productive Partnerships

Establishing productive school-family-community partnerships has become the most commonly embraced policy initiative in schools and school districts (Epstein & Sheldon, 2002). International studies have revealed educators viewing partnerships as central components to learning in Denmark, as making the difference in improving education in Germany, and as exciting opportunities to collaborate and aid students' learning in Scotland (Epstein & Sanders, 1998).

In Asian education, the Program for International Student Assessment (PISA) reveals family influences and mathematics self-concept to be significantly correlated to mathematics achievement (Lemke et al., 2004). The mathematics education community acknowledges the value of such partnerships and encourages teachers to educate parents about new goals and practices in mathematics teaching so that reform efforts can be meaningfully implemented (Price, 1996; Goddard, Tschannen-Moran, & Hoy, 2001).

This chapter involves a discussion of the theory that nurtures productive partnerships and from which stems the model that frames the parental involvement initiatives shared in later chapters of this book.

THEORY OF OVERLAPPING SPHERES OF INFLUENCE

When educators view the family as separate from a child's schooling, an inaccurate message that the family is expected to do its job and leave the education of children to the schools is likely to form. A more accurate message needs to be conveyed that states both the family and the community are partners with the school in children's education and development (Epstein, 1995).

The theory of overlapping spheres of influence proclaims such a message by identifying students as the main actors in their education, supported by family members at home, administrators and teachers at school, and others in their communities who can offer their talents as partners in a well-balanced learning community (Epstein, 1987). Combining the efforts of these supporting factors benefits students, strengthens families, and improves schools.

Since well-planned and well-implemented family involvement activities contribute to student achievement and success in school (Epstein, 2001), a key point of the theory of overlapping spheres of influence is that parents need to become involved from the beginning and stay involved in their child's academic life, rather than just when their children are in immediate need of help.

Epstein and Sanders (1998) report that "good partnerships provide parents with the information they need to remain involved in their children's education, change teachers' attitudes about parents' helpfulness, and allow students to see that their parents care about schoolwork and homework" (p. 27).

When families are viewed as separate and distinct from school and even from each other, it is difficult to understand the perspectives of administrators, teachers, and other families. It is through both verbal and written communication, as well as hands-on experiences with mathematics, that the entire learning community understands and functions.

It is with this thought in mind that the parental involvement initiatives discussed in later chapters of this book focus on strengthening the home-school connection concerning mathematical learning. The initiatives seek to foster a sense of partnership and community by bringing families together to engage in mathematical tasks, reflect upon their experiences, and share the challenges and triumphs they encounter while doing mathematics.

To cultivate such an atmosphere, Epstein and Jansorn (2004) emphasize the need for a welcoming environment that engages families in activities that contribute to students' academic success and positive attitudes. Their research reveals a framework of six types of involvement that can help establish and strengthen a comprehensive program of school, family, and community partnerships.

These types of involvement include parenting, communicating, volunteering, learning at home, decision making, and collaborating with the community. Epstein and Jansorn stress that these forms of involvement are a means of reflecting today's vision where parents and schools are viewed as equal entities of a learning community.

The ways we establish partnerships makes a difference in how parents and how many parents become and stay involved in their children's education (Day, 2000). When schools focus on implementing activities that reflect the different forms of involvement indicated above, schools are providing opportunities for parents to get involved at school and at home in ways that address both student and family needs and take family schedules into consideration (Epstein & Salinas, 2004).

As stated previously, it is from the theory of overlapping spheres of influence and the reported work of Joyce Epstein and her colleagues that provided the theoretical framework for the design of the parental involvement initiatives of this book. Related research on parental involvement, activities developed by mathematics educators, and my own experiences with several school communities contributed to the field-tested finished products that I share with you. It is my hope that you use them as well as design your own by using the underlying model that I describe in this next section.

A MODEL FOR PARENTAL INVOLVEMENT INITIATIVES CONCERNING MATHEMATICS EDUCATION

The model guides the ways you empower parents' ability to facilitate learning at home with respect to mathematics education. It strengthens parenting skills by helping parents create home conditions that support mathematics learning, and it enhances communication by establishing channels of discourse about mathematical learning.

Parents are invited to volunteer themselves as partners in their child's mathematics learning both at school events and at home. The main goal is to nurture collaboration between parents and children that is supportive of one another, reflects National Council of Teachers of Mathematics

(NCTM) reform efforts, and fuels excitement about doing mathematics together.

Each initiative emphasizes the importance of sharing material/activities with parents relevant not only to their child's grade level but also for all grade levels (Pre-K to 8) to promote a global sense of mathematics learning across the grades. For example, third-grade parents involved with an initiative's topic and related tasks are also made aware of the relevancy of the topic and its related tasks in other grades.

We don't want misperceptions forming about methods of teaching, such as the use of manipulatives being for some grades and not others. Rather, the power of the practice, for example, the use of manipulatives, throughout the grades needs to be demonstrated to garner consistent support.

The model consists of an invitation, an initial meeting, an engagement workshop, a home activity, a follow-up session, and additional tasks to maintain connections between the mathematics classroom and the home.

Integrated into the model are features that (1) *inform* the parent community about the importance of their involvement and the rationale behind specific methods used in the mathematics classroom; (2) *engage* both parents and their children in mathematics learning activities at workshop settings and in the home; (3) *promote reflection* as a community of parents and children about their learning experiences; and (4) *maintain connections* between the home and the mathematics classroom.

These features are described in general form at this point and are addressed again in later chapters in which specific initiatives are discussed in depth.

- Inform
  - Epstein and Sanders (1998) report families' attitudes toward school improving when they are invited to participate in their children's education. Guastello (2004) states that, if schools and students reach out and invite parents into a "village of learners," they are encouraged to come. With this rationale in mind, invitations to parents are sent home for each initiative with personal covers designed by their children.
  - This invitation serves to inform parents of the initiative's intent and to request their commitment to participate in an initial workshop, an engagement workshop, a home activity, a follow-up session, and ad-

ditional tasks to maintain connections between the mathematics classroom and the home.
- The initial meeting of each initiative is attended only by parents. It is basically informational and involves parental issues. It commences with a presentation of an agenda for the entire initiative in order to establish direction concerning the initiative's goals.
- To enhance parents' awareness and understanding of best practices in the mathematics classroom, this presentation is followed by a discussion about the changes in mathematics teaching through the years, the constructivist foundation to teaching mathematics, the need for parental involvement, and ways that parents can productively collaborate with their children in reinforcing and extending classroom learning of mathematics to the home.
- To set the stage for the engagement workshop of each initiative, the importance of an initiatives' focus is presented. Supporting research that illuminates the value of an initiative's content is communicated to parents to validate their continued participation in the initiative.

- Engage
  - Social cognition theorists such as Vygotsky state that, through socially meaningful activity, higher mental processes and ideas occur (Kozulin, 1987). Such meaningful activity for parents and their children is nurtured during each initiative by engaging parents in ways that extend beyond checking homework and fostering rote memorization of mathematics facts.
  - Each initiative seeks to involve parents in ways that build awareness of their varied roles in mathematics education. It is critical that they see themselves as agents of mathematics reform who need to partner with their children and their children's teachers.
  - Parents and children are stimulated to learn from each other through active engagement and discussion to build both parents' awareness of their children's mathematical thinking and children's awareness of their parents as collaborators in their learning of mathematics.

- Promote Reflection
  - John Dewey describes a reflective practitioner as someone who has the ability to think critically about their work (Dewey, 1904). This practitioner in turn explores the classroom environment and creates or modifies their own pedagogical practices to establish new and

improved contexts for learning (Schon, 1983). In the same way, a reflective parental partnership is an effective partnership as well.
- To help nurture a community of reflective learners, each initiative allocates time for reflection on learning experiences in both verbal and written forms between parents and their children, among parents, and among children. This allows all partners in the learning community to be informed. Needed modifications at home and/or in the classroom can be targeted and addressed as well.

- Maintain Connections
  - Parents must consistently remain involved in their children's mathematics education, voice concerns, and contribute information (Peressini, 1998) for productive partnerships to continue to thrive. Each initiative maintains such connections between home and school through interactive homework, classroom follow-up strategies, and an interactive newsletter.
  - Homework is only effective if it is relevant, creative, and meaningful to the student and parent (Sullivan & Seequeira, 1996; Walberg, Paschal, & Weinstein, 1985). Therefore, the interactive homework (homework designed in a manner that enables students to share their mathematical thinking and collaborate with parents) helps to maintain quality levels of parental involvement.
  - Sometimes parents and their children quarrel over how to accomplish a mathematics task, and it may end in an unproductive clash of wills. To prevent such a mishap, the interactive homework includes helpful questions that foster deeper thinking of mathematics.
  - Opportunities for parents to share their own strategies and at the same time respect their child's way of doing things (Epstein, 2001) are integrated into the assignments. The interactive homework also provides parents with an opportunity to communicate with the teacher through a parent comment sheet.
  - Classroom follow-up strategies involving both whole class and small group discussions along with written reflections depict significant ways for teachers to monitor parental involvement that goes beyond just checking off who has and has not turned in work they did with their parents. These teaching strategies are discussed in later chapters and give children an additional reason for doing their best at home with their parents.

- An interactive newsletter for each initiative serves as yet another form of communication between the mathematics classroom and the home. The newsletter summarizes the information and learning experiences of the entire initiative as well as provides follow-up materials for home use in such forms as relevant websites, text resources, and children's literature.
- Parents are also provided with another opportunity to communicate with teachers about mathematics learning at home through the newsletter's parent comment sheet.

KEY POINTS TO SHARE WITH PARENTS

Summarized information is presented at this point to serve both as a review of material discussed in this chapter and as organized information to facilitate the initial workshop for each initiative discussed in later chapters of this book. I again suggest using the information as a guided script for yourself and to create transparencies, PowerPoint presentations, and hand-outs so that parents have materials to refer to both during the initiative's initial workshop as well as at home.

As mentioned previously in chapter 1, to avoid repetition in cases where multiple initiatives are conducted with the same group of parents, this information can be shared once and then reviewed on subsequent occasions or adapted according to your individual situations.

Theory of Overlapping Spheres of Influence

- Students are the main actors in their education, supported by others at home, at school, and in their communities.
- Combined efforts benefit students, strengthen families, and improve schools.
- Parents need to become involved from the beginning and stay involved in their child's academic life.

Main Goals of the Parental Involvement Initiative

- To build a community of mathematics learners that support one another by partnering parents with teachers, parents with their children, and parents with other parents.

- Establish mathematics learning communities that reflect NCTM reform efforts and excite both parents and their children about doing mathematics together.
- Open the lines of communication between and among parents and their children, teachers and parents, and parents.

Components of the Initiative

- Initial Meeting (Attended only by parents)
  - Introduction of the initiative's structure
  - Information session on mathematics best practices and content of the initiative
- Engagement Workshop (Attended by parents and children)
  - Parents and children working together and reflecting on mathematical tasks
- Home Activity (Completed by parents and children)
  - Mathematics learning and collaboration extended to the home
- Follow-up Session (Attended by parents and children)
  - Parents and children reflecting on their home mathematics learning experiences
- Additional Tasks (Maintain Connection)
  - Interactive Homework
  - Classroom Follow-up Strategies
  - Interactive Newsletter

SUGGESTIONS FOR FURTHER READING

Carey, L. (1998). Parents as math partners: A successful urban story. *Teaching Children Mathematics*, 4(6), 314–20.

Chadwick, K. G. (2004). *Improving schools through community engagement*. Thousand Oaks, CA: Corwin Press, Inc.

Edge, D. (Ed.). (2000). *Involving families in school mathematics*. Reston, VA: NCTM.

Ensign, J. (1998). Parents, portfolios, and personal mathematics. *Teaching Children Mathematics*, 4(6), 346–72.

Epstein, J. L., Sanders, M. G., Simon, B. S., Salinas, K. C., Jansorn, N. R., & Voorhis, F. L. (2002). *School, family, and community partnerships*. Thousand Oaks, CA: Corwin Press, Inc.

Gilliland, K. (2002). Families ask. *Mathematics Teaching in the Middle School*, 7(9), 510–11.

Kyle, D., McIntyre, E., Miller, K., & Moore, G. (2002). *Reaching out*. Thousand Oakes, CA: Corwin Press, Inc.

Litton, N. (1998). *Getting your message out to parents*. Sausalito, CA: Math Solutions Publications.

Lowry, K. (2002). The world of math online. *Teaching Children Mathematics*, 8(9), 531.

McElwain, D. (2001). Parent partners: Workshops to foster school/home/family partnerships. *Teaching Children Mathematics*, 8(4), 252.

McEwan, E. K. (1998). *How to deal with parents who are angry, troubled, afraid, or just plain crazy*. Thousand Oakes, CA: Corwin Press, Inc.

Pagni, D. (2000). An educator's guide to answering parents' questions on mathematics [review of book]. *Teaching Children Mathematics*, 7(1), 44.

Santora, N. (2000). *Math fun: A guide for teachers, parents, and mentors of grade school children*. Lincoln, NE: iUniverse, Inc.

Silby, R. (1999). Letters to parents in math: 30 Ready-to-use letters in English and Spanish [review of book]. *Teaching Children Mathematics*, 6(4), 270.

Topping, K. (1998). *Parental involvement and peer tutoring in mathematics and science*. London: David Fulton Publishers.

REFERENCES

Benken, B., & Brown, N. (2004). Improving students' mathematical understandings: An exploration into teacher learning in an urban professional development setting. Paper presented at the North American Chapter of the International Group for the Psychology of Mathematics Education, Toronto, Canada.

Bruner, J. (1986). *Actual minds, possible worlds*. Cambridge, MA: Harvard University Press.

Day, C. (2000). Teachers in the twenty-first century: Time to renew the vision. *Teachers and Teaching: Theory and Practice*, 6, 101–15.

De La Cruz, Y. (1999). Reversing the trend: Latino families in real partnerships with schools. *Teaching Children Mathematics*, 5(5), 296–300.

Dewey, J. (1904). The relation of theory to practice in education. In C. McMurray (Ed.), *The relation of theory to practice in the education of teachers: Third yearbook for the National Society of the Scientific Study of Education*. Chicago: University of Chicago Press.

Dienes, Z. *Building up mathematics*. London: Hutchinson Education, 1960.

Drake, C., Spillane, J., & Huffred-Ackles, K. (2001). Storied identities: Teacher learning and subject-matter context. *Journal of Curriculum Studies*, 33(1), 1–23.

Epstein, J. L. (1987). Toward a theory of family-school connections: Teacher practices and parental involvement. In K. Hurrelmann, F. Kaufmann, & F. Losel (Eds.), *Social intervention: Potential and constraints*. New York: DeGruyter, 121–36.

Epstein, J. L. (1995). School/family/community partnerships. *Phi Delta Kappan*, 76(9), 701–13.

Epstein, J. L. (2001). Introduction to the special section: New directions for school, family, and community partnerships in middle and high schools. *NASSP Bulletin*, 85(627), 3–6.

Epstein, J. L., & Jansorn, N. (2004). School, family, and community partnerships link the plan. *Education Digest*, 69(6), 19–23.

Epstein, J. L., & Salinas, K. (2004). Partnering with families and communities. *Educational Leadership*, 61(8), 12–17.

Epstein, J. L., Salinas, K., & Jackson, V. (1995). *Tips (Teachers involve parents in schoolwork)*. Baltimore, MD: Center on School, Family, and Community Partnerships, Johns Hopkins University.

Epstein, J. L., & Sanders, M. (1998). International perspectives on school-family-community partnerships. *Childhood Education*, 74(6), 340–41.

Epstein, J. L., & Sheldon, S. (2002). Present and accounted for: Improving student attendance through family and community involvement. *Journal of Educational Research*, 95(5), 308–20.

Flores, A., & Brittain, C. (2003). Writing to reflect in a mathematics methods course. *Teaching Children Mathematics*, 10(2), 112–18.

Goddard, R., Tschannen-Moran, M., & Hoy, W. (2001). A multilevel examination of the distribution and effects of teacher trust in students and parents in urban elementary schools. *Elementary School Journal*, 102(1), 3–17.

Grant, S. G. (1998). *Reforming reading, writing, and mathematics: Teachers' responses and the prospects for systemic reform*. Mahwah, NJ: Erlbaum.

Guastello, E. F. (2004). A village of learners. *Educational Leadership*, 61(8), 79–83.

Koch, J. (2005). *Science stories*. Boston: Houghton Mifflin Company.

Kozulin, A. (1987). Book reviews: Vygotsky and the social formation of mind. *American Journal of Psychology*, 100(1), 123.

Lemke, M., Sen, A., Pahlke, E., Partelow, L., Miller, D., Williams, T., et al. (2004). *International outcomes of learning in mathematics literacy and problem solving: PISA results from the U.S. perspective*. Washington, D.C.: National Center for Education Statistics.

Loucks-Horsley, S., & Matsumoto, C. (1999). Research on professional development for teachers of mathematics and science: The state of the scene. *School Science and Mathematics*, 99(5), 258–71.

Noddings, N. (1993). Constructivism and caring. In R. D. Davis and C. A. Maher (Eds.), *School mathematics, and the world of reality*. Boston: Allyn & Bacon, 35–50.

Organisation for Economic Co-operation and Development. (2004). *Learning for tomorrow's world: First results from PISA (Program for International Student Assessment) 2003*. Paris: Author.

Peressini, D. (1997). Parental involvement in the reform of mathematics education. *Mathematics Teacher*, 90(6), 421–27.

Peressini, D. (1998). The portrayal of parents in the school mathematics reform literature: Locating the context for parental involvement. *Journal for Research in Mathematics Education*, 29(5), 555–76.

Piaget, J. (1974). *The child and reality: Problems of genetic psychology*. London: Frederick Muller (original work published 1972).

Price, J. (1996). President's report: Building bridges of mathematical understanding for all children. *Mathematics Teacher*, 89(6), 536–39.

Reys, R., Lindquist, M., Lambdin, D., Smith, N., & Suydam, M. (2001). *Helping children learn mathematics*. New York: John Wiley & Sons, Inc.

Schon, D. A. (1983). *The reflective practitioner: How professional think in action*. New York: Basic Books.

Sheldon, S., & Epstein, J. L. (2001). Focus on math achievement: Effects of family and community involvement. Paper presented at the 2001 annual meeting of the American Sociological Association, Anaheim, CA.

Sullivan, M. H., & Seequeira, P. V. (1996). The impact of purposeful homework on learning. *Clearing House*, 69, 346–48.

Van De Walle, J. (2001). *Elementary and middle school mathematics*, 4th ed. New York: Addison Wesley Longman, Inc.

Walberg, H., Paschal, R., & Weinstein, T. (1985). Homework's powerful effects on learning. *Educational Leadership*, 42(7), 76–79.

Chapter Three

Using Hands-on Moveable Objects

Manipulatives are important tools that help our students understand mathematics. Of equal importance are parents who understand the rationale behind using these movable objects so that they can support this aspect of their child's learning. Unfortunately, most of today's parents' mathematics learning experiences do not exemplify in-depth use of manipulatives (Mistretta, 2004).

To cultivate deep understanding of current mathematics teaching practices, careful thought must be put into the ways we inform our parent populations. They not only need clear information about manipulative use and their connections to mathematics standards, but they also need to engage in using these tools in much the same way as their children do in the mathematics classroom. As the Chinese proverb by Confucius states,

- Tell me, I'll forget.
- Show me, I'll remember.
- Involve me, I'll understand.

To foster such meaningful understanding about the use of manipulatives in mathematics learning, this chapter describes a parental involvement initiative designed to enhance parents' awareness of and confidence in using the manipulative known as tangrams. This parental involvement initiative, as well as the initiatives presented in chapters 4 and 5, can be facilitated by yourself with your own class, or with multiple classes where you co-present with your colleagues.

The initiative's components (an invitation to parents, guidelines and content information for an initial meeting and an engagement workshop, a home activity, a follow-up session format, an interactive homework assignment, classroom follow-up strategies, and an interactive newsletter) are discussed in detail and shared as sequenced action steps toward cultivating productive collaboration between parents and their children concerning mathematical learning.

ENCOURAGE YOUR PARENTS TO COME

As stated in chapter 1, if schools and students reach out and invite parents into a "village of learners," they are encouraged to participate in events (Guastello, 2004). Sending an invitation that informs parents of an initiative's intent, requests their commitment to participate, and inquires about times that would best suit their busy schedules creates a feeling of consideration and real partnership (De La Cruz, 1999).

Responses can then be reviewed and the most appropriate meeting times planned. To personalize the experience, students should design their own cover for their parents' invitation. This can be done by folding a piece of construction paper downward in half and having the children design the front portion of it. The invitation can then be folded in the same way and placed inside the cover forming a booklet.

I have found this task to be exciting for children because they are given the opportunity to put their personal touch on an event that is being planned for both them and their parents. Parents tend to take more notice as well because their child is coming home with an invitation that both they and the school created. They see their children as well as the school reaching out to them.

The text of the invitation (letter of intent and inquiry) informs parents of the initiative's intent, agenda, and possible meeting times. It is presented at this point for you to use when crafting your own letter of intent and inquiry or amend according to your needs. Also included are a response form and a letter of announcement.

Letter of Intent and Inquiry

Dear Parents,
Children need to find mathematics experiences interesting and meaningful to reach their full mathematics potential. To do this, using manipula-

tives (moveable objects) is a teaching method that has proven to be successful in developing students' ability to understand mathematics concepts, master skills, and problem solve.

Your involvement with such a practice plays a vital role in supporting your child's mathematics education. To empower you as partners with this aspect of your child's education, a parental involvement initiative titled "Using Hands-on Moveable Objects" has been designed for you and your child.

This initiative involves an initial meeting (1 hour), an engagement workshop (1 1/2 hours), a home activity, a follow-up session (1 hour), and additional tasks to help maintain connections between the mathematics classroom and your home. The initiative *informs* you about the importance of parental involvement and the rationale behind specific teaching methods used in today's mathematics classrooms.

The initiative also *engages* you and your child in activities involving the use of moveable objects known as tangrams in a workshop setting that extends into your home. Time for *reflection* is facilitated as well between you and your child, among the parent community, and among the student community.

Please be aware that you are expected to attend the initial meeting and engagement workshop, engage in a home activity with your child, attend the follow-up session, and maintain connections between the mathematics classroom and your home. Three meetings at school are required where one is attended only by parents and two are attended by both parent and child.

If you wish to commit to this parental involvement initiative, please return the attached response form. If you have more than one child, please arrange for each child to be represented by one family member.

We want to schedule this initiative in consideration of the busy schedules of all those involved. Therefore, please indicate on the attached form the time that would best suit you.

I look forward to working with you.

Sincerely,

The text of the response form that is mentioned in the above letter of intent and inquiry is presented below. It fosters a sense of commitment and serves to investigate the most appropriate meeting times for your parent population. Parents are given the opportunity to indicate reasons they may not be able to participate as well. Knowing such reasons allows you to make accommodations if possible. We can't assume that parents' lack of participation is due to lack of interest. There may be a very good reason that when known can be remedied.

The response form also serves as an organizational tool. The names of both children and their attending family members need to be indicated on the form to help you keep accurate records of attendance throughout the initiative.

Following the text of the response form is text for a letter of announcement that serves to begin the initiative. Spaces are provided for specific dates for the initial meeting, engagement workshop, and follow-up session. As mentioned previously, the text is shared with you for your direct use in creating your own response form and letter of announcement or as a framework to amend according to your specific situations.

Response Form

Please return this form by _____.
We (commit, cannot commit) to the parental involvement initiative titled "Using Hands-on Moveable Objects."
If you cannot commit to this initiative, please explain why on the back of this form so we can try to partner with you in another way.
Name of Parent(s) and Other Family Members Committed to the Initiative (Each child must be accompanied by one family member.)
_____
_____

Name of Child(ren) (with grade level) Committed to the Initiative
(Indicate the family member above to partner with each child indicated below.)
_____
_____

Parent(s) Signature(s): _____
_____

Student(s) Signature(s): _____
_____
_____

Circle the most appropriate time for your family to commit to this initiative.
Weeknight beginning at 7:00 P.M. (Specify night) _____
Weeknight beginning at 8:00 P.M. (Specify night) _____
Saturday morning beginning at 10:00 A.M. _____
Saturday afternoon beginning at 1:00 P.M. _____
Sunday morning beginning at 10:00 A.M. _____
Sunday afternoon beginning at 1:00 P.M. _____

Letter of Announcement

Dear Parents/Family Members,
To empower you as partners in your child's learning of mathematics, the parental involvement initiative titled "Using Hands-on Moveable Objects" has been planned for you and your child. Your feedback concerning the most appropriate days and times for this event have been carefully considered, and the following dates and times reflect common responses.
Initial Meeting_____
(1 hour session for parents only)
Engagement Workshop_____
(1 1/2 hour session for parents and children)
Follow-up Session _____
(1 hour session for parents and children)

As noted in previous correspondence, the initiative *informs* you about the importance of parental involvement and the rationale behind specific teaching methods used in today's mathematics classrooms. The initiative *engages* you and your child in activities involving the use of tangrams in a workshop setting that extends into your home. Time for *reflection* on your learning experiences is facilitated between you and your child, among the parent community, and among the student community as well.

Please be aware that you have committed to this initiative and are expected to attend the initial meeting and engagement workshop, engage in a home activity with your child, attend the follow-up session, and maintain connections between the mathematics classroom and the home. All meetings start exactly on time, so please be punctual.

I look forward to working with you.

Sincerely,

INITIAL MEETING

The initial meeting should only be attended by parents due to its adult focused content. You should begin the meeting by presenting the initiative's goals and components using the information provided in chapter 2 (Key Points to Share with Parents). This introduction allows parents to become knowledgeable of the initiative's structure and features that serve to inform, engage, promote reflection, and maintain connections between the mathematics classroom and the home.

After this introduction, a discussion can follow with the use of the organized information supplied in chapter 1 (Key Points to Share with Parents) about the changes in mathematics teaching through the years, a constructivist foundation to teaching mathematics, the value of parental involvement, and the ways that parents can productively partner with their child at home to support mathematics education reform efforts.

The discussion can then turn to the topic of using manipulatives, specifically tangrams, in mathematical learning. Helpful information is organized below to facilitate your discussion about the value of using manipulatives in the mathematics classroom as well as some of the specific uses of tangrams. As with the information you share from chapters 1 and 2, I suggest incorporating the following information into transparencies, PowerPoint presentations, and hand-outs.

Manipulatives

- make abstract concepts concrete
- are useful for solving problems
- nurture meaningful understanding of standard mathematical vocabulary and symbolism
- build students' confidence with opportunities to create physical evidence of thinking and reasoning

Tangrams

- develop spatial sense and reasoning skills
- facilitate exploration of geometric shapes and fractional relationships

Present a transparency tangram set to parents using an overhead projector and inform them that tangrams are often used in mathematics classrooms at all grade levels. A transparency set of tangrams may be purchased from ETA/Cuisenaire Company (www.etacuisenaire.com). It is important they not only become aware of their existence but also know how to use them productively at home with their children.

Let the parents know they'll become engaged in activities using tangrams with their child in much the same way that their child experiences them in the mathematics classroom. Share with parents that they will become aware of activities relevant not only to their child's grade level but to other levels as well.

This is important so that the role of tangrams throughout the grades is understood. We don't want a misperception to form that manipulatives are only for the younger grades. Rather, we need to build parents' awareness of the incremental power of these tools throughout the grades.

Allow a question-and-answer session to follow so that parents can contribute feedback and clarify any existing concerns. If questions from a small number of parents require in-depth responses and do not pertain to all those present, speak with these select parents after the meeting to avoid detaining others.

Before closing, remind parents that the engagement workshop involving both them and their children takes place two weeks later. I've found it helpful to send a reminder home to parents a day or two before this event. A two-week interval between the initial meeting and engagement workshop gives parents enough time to review the material presented at the initial meeting and remain vibrant about the initiative's intentions.

ENGAGEMENT WORKSHOP

You are at a point now where you have given your parents the background information they need to understand the rationale behind current mathematics teaching methods and the value of their involvement in their child's mathematics education. The engagement workshop can in turn take on a different environment that is quite active and insightful for parents and their children, and even you, as you explore, communicate, and strategize together about mathematics using tangrams.

To create an environment conducive to such activity, parents and children should sit at tables so that concrete explorations in a group setting are feasible. To build a historical perspective to these learning tools, inform the group that tangrams originated in China in the 1800s and have been used extensively to investigate mathematical ideas.

Distribute the tangram sets and let parents and children know that they are about to become engaged in activities involving spatial reasoning, computational skills, and problem solving.

The purpose of this type of experience should be conveyed to parents as a way to actively and concretely participate in a learning environment typical of their child's everyday mathematics classroom. A tangram set can be found online at http://illuminations.nctm.org/Lessons/DevelopGeometric/Tangram-AS-SlideFlipTurn.pdf. The actual set is located on the second page of this online document.

You can print this tangram set onto cardstock and cut them out for use during the engagement workshop if plastic sets are not available. A tangram set needs to be supplied to each child and their parent for use during this engagement workshop. You should also place two to three additional sets on the tables so that extra pieces are there when needed.

Whether or not you use paper tangrams at the engagement workshop, I suggest making paper tangram sets on cardstock so that each family is supplied with a durable set to use at home since the home activity that families later go home with requires them to use tangrams.

Remember, supplying families with all the necessary tools allows them to easily collaborate at home about mathematics. An alternative to creating all of the paper cardstock sets yourself is to distribute the tangram set on a piece of cardstock to each child and their parent at the end of the engagement workshop. Ask them to cut out the seven pieces themselves. I'll leave this decision up to you as you prepare for this event.

It is important for me at this point to address a challenge that may present itself to you when creating paper tangrams. There are instances when a website that has functioned in the past becomes nonexistent for various reasons. If you find yourself in a situation where the website I mentioned earlier is not functioning, I recommend two alternatives.

The first is to purchase an inexpensive plastic tangram template from ETA/Cuisenaire Company (etacuisenaire.com). Using this template, you can draw the tangram pieces onto a piece of paper and then copy them onto cardstock. The other alternative is to enlarge figure 3.2 that appears later in this section. You can then copy that outlined figure consisting of the seven tangram pieces onto cardstock.

Getting back to the actual format of the engagement workshop itself, a period of time (approximately fifteen minutes) should be provided for free exploration to both foster familiarity with the pieces and satisfy curiosity about them. Usually when a new manipulative is introduced to children in the classroom, a full class period is allotted for free exploration. But since tangrams should be familiar to the children from their experiences in your classroom, and due to time constraints, a shorter amount of time can be spent exploring the pieces during this part of the workshop.

Information about the size and shape of the seven tangram pieces should be elicited from the group (two small triangles, one medium triangle, two large triangles, one square, and one parallelogram) as it would with your students in the mathematics classroom. To guide the group's inquiry and provide a basis for communication between the parents and children, pose the following questions about the relationships among the pieces:

- How does the small triangle compare with the medium triangle?
- How does the small triangle compare with the large triangle?
- How does the medium triangle compare with the large triangle?
- What tangram pieces can be joined together to form other tangram pieces?
- How many ways can you cover the large triangle with other tangram pieces?

Instruct the group to write about and/or draw their discoveries on a sheet of paper that you distribute to them to help as they process their thoughts. This task serves to organize a record of their findings to share later on in the workshop. Circulate the group to give assistance and observe interactions and conversations between the children and their parents.

It can be quite a learning experience for you as well as you observe the ways the children and their parents interact. It is through such on-site observations that you can gain insight into parent-child collaboration concerning mathematical learning and address areas in need of attention.

Conceptual understanding is enhanced when we are given opportunities to concretely explore mathematics and share both discoveries and different ways of thinking about mathematical ideas (Burns, 1996). With this in mind, I encourage you at this point to bring the whole group together to discuss (approximately fifteen minutes) what they discovered about their tangram sets.

You should record their observations on a chart or overhead transparency so that all can see what was learned by the entire group. Engaging parents in this way allows them to see the dynamics of constructivist learning where knowledge is actively obtained by the learner as opposed to passive environments where information is just given by the teacher as off a silver platter and then later memorized.

I have found this discussion to be quite eye-opening for parents that goes beyond acknowledgment of the different shapes involved in a tangram set. The parents I have worked with were pleased with the opportunity to observe their children using fractional language as they responded to the previously stated guiding questions with answers such as follows: "The small triangle is half the size of the medium triangle"; "The small triangle is one-fourth the size of the large triangle"; and "The medium triangle is half the size of the large triangle."

The parents and children also noted how the two small triangles can form both a square and a parallelogram shedding light on the fact that both shapes have the same area because they both contain the same amount of space (the two same sized small triangles) but just in different representations. Conceptual understanding of finding a shape's area by partitioning it into smaller shapes also develops as parents and children explore the answer to the last guiding question that asks the following: How many ways can you cover the large triangle with other tangram pieces?

By covering the large triangle in different ways and communicating about multiple methods of solution, parents and children see how the large triangle can be formed with four small triangles, two medium triangles, the two small triangles and the medium triangle, the two small triangles and the square, and the two small triangles and the parallelogram.

To initiate an activity (approximately twenty minutes) that involves spatial reasoning and connects newly developed knowledge with a computational task, have the group arrange the seven tangram pieces into an outlined cat (see figure 3.1). To ease the cat's formation in grades Pre-K to 2, use an outlined cat with the shapes drawn inside (see figure 3.2).

Figure 3.1.

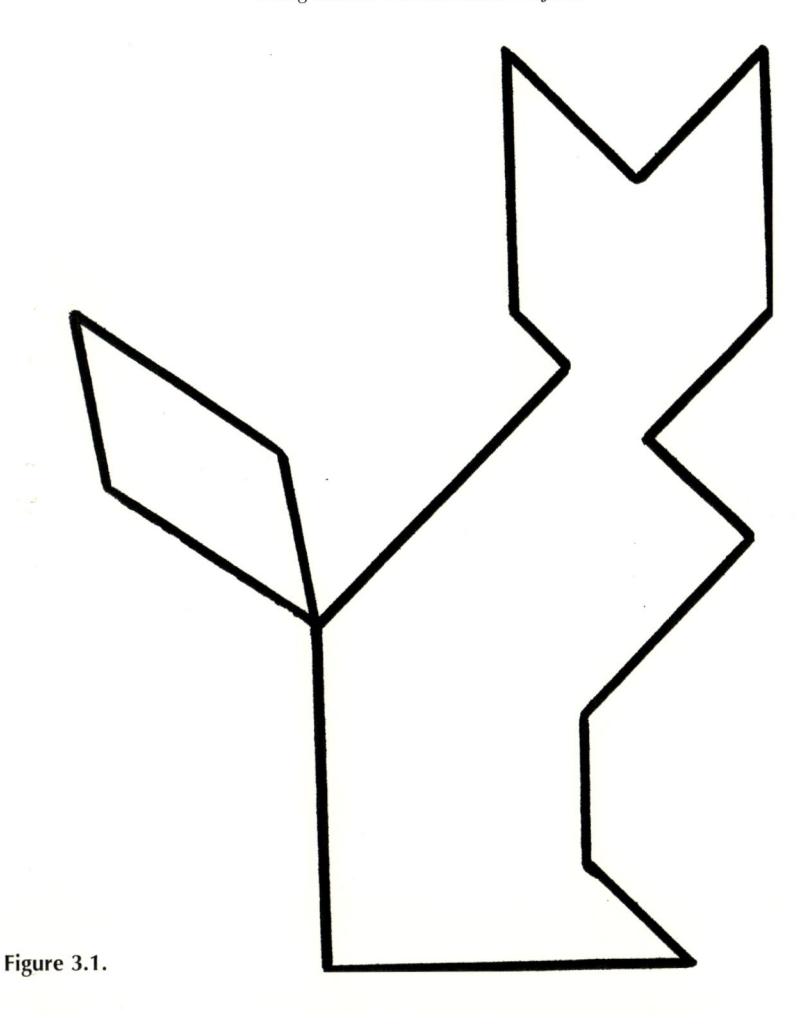

These cats and following related mathematical tasks have been adapted from field-tested activities designed by mathematics educators (Fuys & Tishler, 1979; ETA/Cuisenaire, n.d.). Other outlines are available online at the National Council of Teachers of Mathematics (NCTM) website. The specific address is as follows: http://standards.nctm.org/document/eexamples/chap4/4.4/standalone1.htm.

If you choose to use these other outlines, please know that when viewing them at the NCTM website, the tangram pieces used with these outlines are much smaller than both the pieces of traditional plastic sets and the template I previously suggested you use for making paper tangram sets.

Figure 3.2.

Therefore, you need to create the outline supplied by the NCTM website by arranging and tracing the tangram pieces you are using with the parents and children in the same manner as they are arranged on the NCTM website (click on the "Hint" given on the webpage to see how the tangram pieces are to be placed to form the outline).

The activities that follow use the cat outlines of figures 3.1 and 3.2 above. Assign a monetary value to the smallest triangle of the tangram set and pose the following questions according to grade level:

- Grades Pre-K to 2
- How much does the cat cost if the smallest triangle costs one cent?
- Grades 3 to 5
- How much does the cat cost if the smallest triangle costs twenty cents?
- Grades 6 to 8
- How much does the cat cost if the smallest triangle costs $3.25?

Instruct parents and children to use what they discovered about the relationships among the tangram pieces to arrive at their solutions. Place emphasis on communicating in both written and verbal form so that information and ideas can be gathered and organized, as well as thoughts clarified in a meaningful manner. Advise parents not to do all of the telling. Encourage them to investigate their child's mathematical thinking by asking some or all of the following questions:

- Where shall we begin?
- What do we know that can help us?
- Can we approach this another way?
- Why?
- How?

At this point in the workshop, I have found it best to break (approximately ten minutes) to allow parents and children to stretch and partake in some light refreshments that you provide. You may also find that the workshop has taken more time than expected or group dynamics indicate that enough time has been spent on task. In either case, you may in turn decide to divide this workshop into two parts.

If so, work already completed should be collected. Invite the parents and children back no longer than a week later to complete the workshop. Waiting longer than a week can impinge on the participants' enthusiasm and memory of specific accounts of this event.

My experiences have presented me with situations where it was best to divide this workshop in half. On other occasions, the group's energy and concentrated efforts called for the workshop to be completed in its entirety in one session, even though it went over 1 1/2 hours. I'll leave this decision up to you and your specific situations.

Whether you choose to continue or to segment the workshop, the next step is to allocate time (approximately fifteen minutes) for the group to reflect on their work and share ideas about how they obtained their solutions. Divide the participants into small groups of two to three families and ask them to discuss the following reflection questions:

- What was your solution, and how did you arrive at it?
- Did you and your child approach the problem in the same manner?
- Did you help each other?
- Was this experience challenging?
- Was this experience fun?
- Would you change anything the next time concerning your method of solution or how you worked together?

These reflection questions should be posted on chart paper or a transparency that is visible to the entire group. You may also opt to create question sheets to duplicate and distribute to each of the small groups. In either case, have someone record their responses onto chart paper or a transparency and designate another to later share their responses with the whole group.

Once you notice that the majority of small groups have answered the reflection questions, bring the families back together as a whole group to share their thoughts and listen to the solutions/experiences of others (approximately fifteen minutes). Request that the person designated to present the responses come and lead the conversation about their group's findings. Seats should be arranged so that the other members of the group can come up as well and contribute to the discussion. To promote rich dialogue, I have found it helpful to ask probing questions such as follows:

- Whose method of solution did you use (parent or child)?
- How were your methods of solution developed?
- Were alternate methods considered?

As stated earlier in this chapter, the parental involvement initiatives of this book can be facilitated with your own class or with multiple classes where you co-present with your colleagues. In either situation, please remember, if you are working only with parents and children of one grade

Using Hands-on Moveable Objects   49

level or interval (Pre-K to 2 or 3 to 5 or 6 to 8), you must also discuss with them at this point the questions and possible methods of solution that would be part of other grade level experiences as well.

As mentioned previously, we want parents to understand the power of the tools we use in the mathematics classroom throughout the grades. To facilitate this type of discussion, samples of student work from other grade levels (see figures 3.3 and 3.4) can be shown to the group. Please

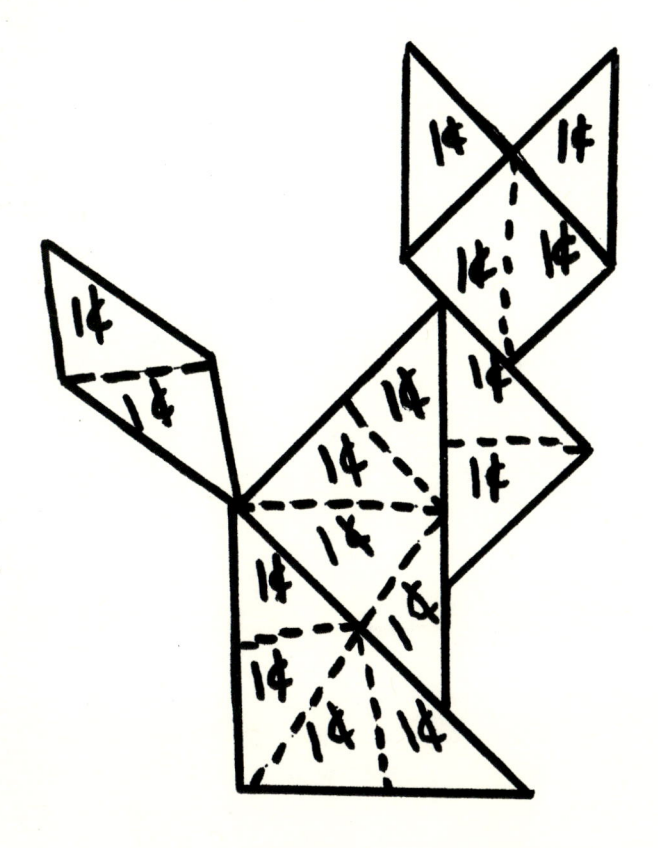

The total is __16¢__

I added everything up.

Figure 3.3.

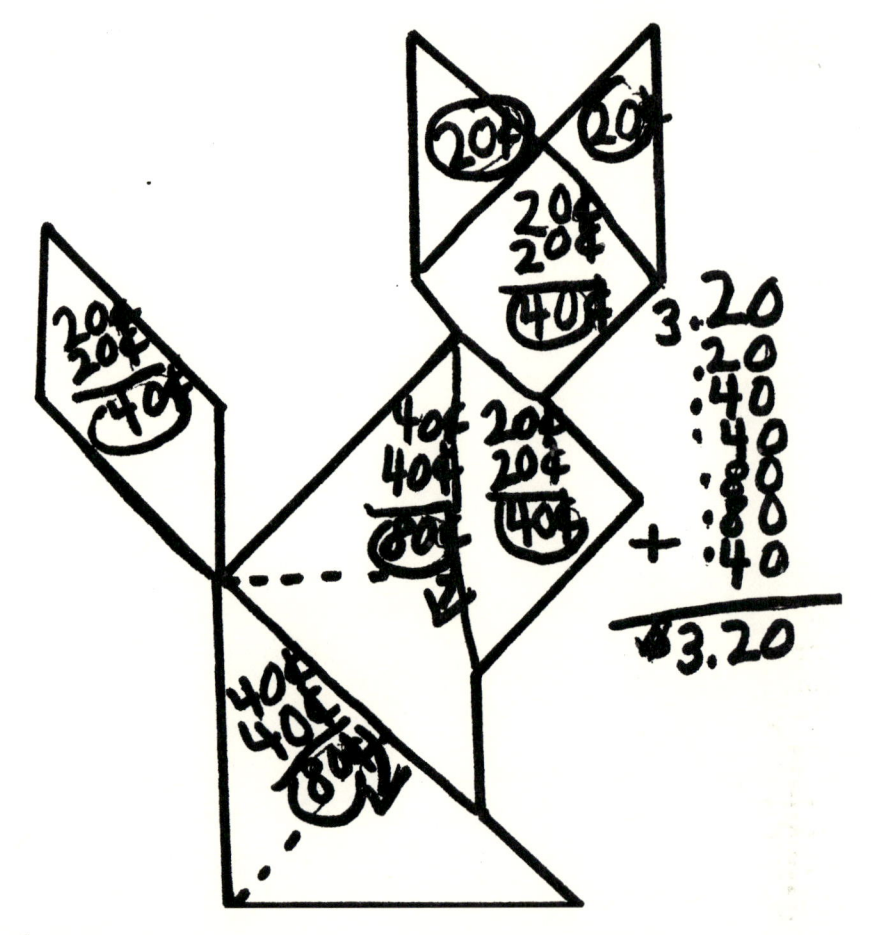

Figure 3.4.

note that the method of solution for the previously stated question for grades 6 to 8 is the same as the one for grades 3 to 5. The numbers used for grades 6 to 8 just require more involved computational skills.

On many occasions, this experience has proven to be as insightful for me as it has for parents and children. Parents have often voiced their appreciation of the concrete experience. They valued how the tangram pieces "help the numbers on paper make sense" and "open their eyes to different ways of approaching a mathematics problem."

Comments such as "It was nice to have time to talk"; "I felt like I got inside my child's head"; "I have a new appreciation for the reasoning that goes on when solving mathematics problems"; and "I realize now that doing mathematics is no longer my way or the highway" have been part of my reflection sessions with parents and children.

The children showed equal levels of enthusiasm. They expressed gratitude for having their parents "come into their world" and experience the way they learn mathematics. Comments have included the following: "Working with my parent was fun"; "It was nice sharing our ideas"; "I listened to my parent's thinking, and she listened to mine"; and "This feels good because now when I go home my parent knows what I'm talking about."

To both reinforce and extend this workshop experience, the stage now needs to be set for learning at home. Text for grade level cover letters, home activities, and helpful content information for parents that reinforce and extend the engagement workshop experience are provided at this point to help facilitate a home learning experience. As mentioned previously, the text is shared with you on the following pages for your direct use or as a framework to amend according to your specific situations.

After the activity is complete, parents are to complete a reflection paper concerning the experience. This reflection paper is included for your use on the following pages as well. Inform the parents that this reflection paper and their responses must be brought to the follow-up session so that discussion as a community of mathematics learners using tangrams can be facilitated.

In closing the engagement workshop, distribute the paper tangrams sets and materials (cover letter, home activity, content information, reflection paper). Discuss the materials with everyone so that the parents and children are clear about what needs to be done as they engage in doing mathematics together at home. Inform the parents that they need to bring all of the work they do with their children at home to the follow-up session.

Emphasize the need to collaborate with their children in the same manner they did during the engagement workshop. Remind them not to do all of the telling but rather to communicate and explore thought processes together. Most important, emphasize the importance of just having fun at home with mathematics.

Cover Letter for Grades Pre-K to 2

Dear Parents and Students,

Presented to you in this letter is a mathematics activity to do at home that both reinforces and extends our learning experiences from the engagement workshop. The first part of the activity is similar to what we did together. You need to show your thinking while you solve the problem with pictures and/or words.

The second part asks you to think about what you could do to make your design cost more, and what you could do to make your design cost less. Share your thinking about how you arrived at your answers with pictures and/or words. Parents, remember that if you do all of the telling this is not a productive collaboration. To optimally communicate and explore thought processes, you should pose some or all of the questions you did during the engagement workshop. These questions include the following:

- Where shall we begin?
- What do we know that can help us?
- Can we approach this another way?
- Why?
- How?

Our follow-up session takes place on _____ at _____ so we can discuss your work and experiences. This session (approximately one hour) starts exactly on time, so please be punctual. I look forward to seeing your solutions and discussing with you your methods of solution and learning experiences. As you work together on this activity, keep in mind the importance of exploring mathematics together, sharing your thinking, and just having fun.

Sincerely,

Home Activity for Grades Pre-K to 2

Take a blank piece of paper and design your own shape onto it using your paper tangrams. Paste your design onto the paper. Find out how much the entire design costs if the smallest triangle costs three cents. Show your thinking on your piece of paper. On a separate piece of paper, answer the following questions:

- What can you do so that your shape costs more?
- What can you do so that your shape costs less?

Content Information for Grades Pre-K to 2

Dear Parents,

As you and your child explore this mathematics activity together, engage in conversation about your child's thinking as well as your own. Use the guiding questions included in the cover letter to facilitate your discussion.

The two questions at the end of the activity are asking you and your child to reason about how to raise and then lower the cost of your design. The goal is to have your child determine that the cost of the smallest triangle of the tangram set affects the cost of the total design. If the cost of this piece increases, the total cost increases as well and vice versa. Have your child use actual monetary values, and think this through by posing questions such as follows:

- How did we figure out the total cost of our design at the engagement workshop?
- Did the value of any particular piece have anything to do with the cost of the other pieces? How?
- If we change the cost of a particular piece, does that change the cost of the entire design? How?

This list of questions is meant to help guide your discussion. If other questions that you think of seem more helpful in your specific situation, please be confident enough to use them.

I hope you enjoy your quality math time together!

Sincerely,

Cover Letter for Grades 3 to 5

Dear Parents and Students,

Presented to you in this letter is a mathematics activity to do at home that both reinforces and extends our learning from the engagement workshop. The first part of the activity is similar to what we did together. You need to show your thinking while you solve the problem with pictures and/or words.

The second part asks you to think about what you could do to make your design cost more than one dollar but not more than two dollars. Share your thinking about how you arrive at your answers. Parents, remember that if you do all of the telling this is not a productive collaboration. To optimally communicate and explore thought processes, you should pose some or all of the questions you did during the engagement workshop. These questions include the following:

- Where shall we begin?
- What do we know that can help us?
- Can we approach this another way?
- Why?
- How?

Our follow-up session takes place on _____ at _____ so we can discuss your work and experiences. This session (approximately one hour) starts exactly on time, so please be punctual. I look forward to seeing your solutions and discussing with you your methods of solution and learning experiences. As you work together on this activity, keep in mind the importance of exploring mathematics together, sharing your thinking, and just having fun.

Sincerely,

Home Activity for Grades 3 to 5

Take a blank piece of paper and design your own shape onto it using your paper tangrams. Paste your design onto the paper. Find out how much the entire design costs if the smallest triangle costs fifty-three cents? Show your thinking on the piece of paper. On a separate piece of paper, answer the following question:

- What can you do so that the cost of your shape increases by more than one dollar but not more than two dollars?

Content Information for Grades 3 to 5

Dear Parents,

As you and your child explore this mathematics activity, engage in conversation about your child's thinking as well as your own. Use the guiding questions included in the cover letter to facilitate your discussion. The question at the end of the activity is asking you to reason about how to raise the cost of your design by more than one dollar but not more than two dollars. The goal is to have your child determine that the cost of the smallest triangle of the tangram set affects the cost of the total design. If the cost of this piece increases, the total cost increases as well and vice versa.

Experimenting with higher monetary values for the smallest triangle, computing the corresponding costs of the other tangram pieces, and determining the total cost of the design helps your child determine ways to exceed the original total cost by more than one dollar but not more than two dollars. Help your child think this through by posing questions such as follows:

- How did we figure out the total cost of our design at the engagement workshop?
- Did the value of any particular piece have anything to do with the cost of the other pieces? How?
- If we change the cost of a particular piece, does that change the cost of the entire design? How?
- What plan should we use to find a solution to this problem?
- Is there more than one solution?

This list of questions is meant to guide your discussion. If other questions that you think of seem more helpful in your specific situation, please be confident enough to use them.

I hope you enjoy your quality math time together!

Sincerely,

Cover Letter for Grades 6 to 8

Dear Parents and Students,
Attached is a mathematics activity for you to do that both reinforces and extends our learning from the engagement workshop. The first part of the activity is similar to what we did together. You need to show your thinking while you solve the problem.

The second question asks you to think about what you could do to make the cost of your design increase by more than 10 percent but not more than 15 percent. Share your thinking about how you arrived at your answers. Parents, remember, if you do all of the telling this is not a productive collaboration. To optimally communicate and explore thought processes, you should pose some or all of the questions you did during the engagement workshop. These questions include the following:

- Where shall we begin?
- What do we know that can help us?
- Can we approach this another way?
- Why?
- How?

Our follow-up session takes place on _____ at _____ so we can discuss your work and experiences. This session (approximately one hour) starts exactly on time, so please be punctual. I look forward to seeing your solutions and discussing with you your methods of solution and learning experiences. As you work together on this activity, keep in mind the importance of exploring mathematics together, sharing your thinking, and just having fun.

Sincerely,

Home Activity for Grades 6 to 8

Take a blank piece of paper and design your own shape onto it using your paper tangrams. Paste your design onto the paper. Find out how much the entire design costs if the smallest triangle costs $8.95? Show your thinking on your piece of paper. On a separate piece of paper, answer the following question:

- What can you do so that the cost of your shape increases by more than 10 percent but not more than 15 percent?

Content Information for Grades 6 to 8

Dear Parents,

As you and your child explore this mathematics activity, engage in conversation about your child's thinking as well as your own. Use the guiding questions included in the cover letter to facilitate your discussions.

The question at the end of the activity is asking you to reason about how to raise the cost of your design by more than 10 percent but not more than 15 percent. The goal is to have your child determine that the cost of the smallest triangle of the tangram set affects the cost of the total design. If the cost of this piece increases, the total cost increases as well and vice versa. The formula for determining the percent of increase is as follows:

Percent of Increase = (New Cost − Original Cost)/Original Cost x 100

Experimenting with higher monetary values for the smallest triangle, computing the corresponding costs of the other tangram pieces, determining the total cost, and finding the percent of increase (using the formula) helps your child determine ways to exceed the original total cost by more than 10 percent but not more than 15 percent. Help your child think this through by posing questions such as follows:

- How did we figure out the total cost of our design at the engagement workshop?
- Did the value of any particular piece have anything to do with the cost of the other pieces? How?
- If we change the cost of a particular piece, does that change the cost of the entire design? How?
- How can we determine the percent of increase?
- What plan should we use to find a solution to this problem?
- Is there more than one solution?

This list of questions is meant to help guide your discussion. If other questions that you think of seem more helpful in your specific situation, please be confident enough to use them.

I hope you enjoy your quality math time together!

Sincerely,

Reflection Paper

Bring your written responses to the following questions with you to our follow-up session.

- How were the activities most helpful?
- What did you find most interesting about using the tangrams?
- Did any surprises occur?
- How has this experience impacted your family?

FOLLOW-UP SESSION

The main goal of this session is to share families' work and reflect on the experience as a community of learners. Allow two weeks before holding the follow-up session so that parents have enough time to complete the activities with their child in a manner that is not rushed. I've found that sending a reminder letter home a week before gives parents a heads up to this upcoming event.

Begin the session by placing families into the same smaller groups they were in during the engagement workshop. This works well since there is already an established sense of familiarity among the group members. Instruct parents and children to share their progress with the group concerning the work they did together at home (approximately twenty minutes).

They should use the reflection paper questions and their responses to guide their discussions. Post these questions on chart paper or a transparency for all to see, and designate two people as you did during the engagement workshop to record responses and report on them.

When you notice that all has been accomplished within the groups, bring the entire group together for a discussion (approximately twenty minutes) and record their responses on chart paper or a transparency for all to see. Past experience has informed me that such reflection empowers parents to support each other as a community of mathematics learners.

Some parent comments that I have gathered include the following: "It's now easier for me to assist my child in a nonthreatening way"; "This initiative helped me connect with my child by making learning fun"; "I now know my child's way of thinking better and what is expected"; and "This initiative helped me spend quality time with my child and enjoy the different ways of thinking about mathematics."

During this follow-up session, parents have often expressed how surprised they were to discover how much their children could do. Comments included the following: "I hadn't taken the time to listen to my child think. Now I listen and guide rather than insist on using my way to solve a problem." On many occasions, parents have expressed gratitude for the opportunity to come together as a community of learners. Many commented that the communication between them and their children as well as among fellow parents helped "demystify" mathematics.

Take the remaining time to share this initiative's connections to the *Principles and Standards for School Mathematics* (NCTM, 2000) and highlight the action steps that serve to maintain the home-school connection, namely, the interactive homework assignment, classroom follow-up techniques, and the interactive newsletter.

It is important to highlight for parents how the tasks they engaged in connected the three essential components of mathematics instruction: conceptual understanding, computational skill, and problem solving. Inform parents how they along with their children strengthened their conceptual understanding of two-dimensional shapes when they explored and compared the tangram pieces (Geometry Standard).

They discussed the number of pieces in the set along with commonly used fractions to represent relationships among them (specifically 1/2). As they visualized and used spatial reasoning to design the cat, they also immersed in computational skill building when they computed the total cost of the cat as well as their own designs (Number and Operations Standard).

Higher order thinking skills permeated the tasks as well as they reasoned about how to place the shapes, determined methods of solution, and brainstormed about ways to increase the total cost of their designs (Problem-Solving Standard).

Verbal and written interaction was present (Communication Standard) as participants discussed and recorded their mathematical thinking as well as considered the mathematical thinking and strategies of others.

These multiple connections to the mathematics curriculum as discussed in the *Principles and Standards for School Mathematics* (NCTM, 2000) are summarized below to help you build parents' awareness of the many facets of mathematical learning that they were exposed to during this initiative. I suggest using this summarized information when creating presentation materials such as transparencies, PowerPoint slides, and/or handouts to be used during this closing segment of the follow-up session.

- Geometry
  - Recognize and compare two-dimensional shapes
  - Investigate the results of putting together two dimensional shapes
  - Use visualization and spatial reasoning
- Number and Operations
  - Understand "how many" in a set
  - Understand and represent commonly used fractions
  - Compute efficiently
- Problem Solving
  - Solve problems that arise in mathematics
  - Apply and adapt a variety of appropriate strategies to solve problems
- Communication
  - Communicate mathematical thinking coherently and clearly to peers, teachers, and others
  - Analyze and evaluate the mathematical thinking and strategies of others
- Connections
  - Understand how mathematical ideas interconnect and build on one another to produce a coherent whole

The interactive homework assignment, classroom follow-up techniques, and interactive newsletter are outlined at this point for your presentation purposes as well. These action steps for maintaining the home-school connection established through this initiative are discussed in detail in the following section of this chapter.

- Interactive Homework
  - Engages both child and parent in a mathematics task where input from both parties creates the final product
  - Gives parents the opportunity to communicate with their child about their reasoning
  - Gives parents the opportunity to communicate with their child's teacher about the progress made with the assignment and any existing concerns

- Classroom Follow-up Techniques
  - Verbal and Written Communication
  - Group Discussion
  - Written Reflection
- The above techniques allow the students to
  - reinforce conceptual understanding and skill
  - interact as a community of learners who are growing with their parents as partners in mathematics learning
  - organize their thoughts and give insight into what has been successful and what needs to be improved
- Interactive Newsletter
  - Reviews the events of the initiative
  - Provides additional support for learning at home (websites, children's books, additional resources)
  - Shares classroom happenings
  - Opens lines of communication with parents to an even greater extent.

MAINTAINING THE CONNECTION

To ensure continuity of the strides you have made during this initiative, your students and their parents need to remain connected. In this section, I share ways to accomplish this through an interactive homework assignment, classroom follow-up techniques, and an interactive newsletter designed to keep the home-school mathematics connection alive.

Interactive Homework

Interactive homework engages both the child and their parent in a mathematics task where input from both parties creates the final product. It gives parents the opportunity to communicate not only with their child about their reasoning but also with their child's teacher about the progress made with the assignment and any existing concerns.

The interactive homework assignment shared below engages parents and children in using the attributes (shape and area) of the tangram pieces that they used during the engagement workshop.

The format has been adapted from the work of Epstein and Van Voorhis (2002) and consists of (1) a "Dear Math Learning Partner" section where the parent is invited to work with their child on a mathematics task; (2) a "Take a Look" section to give necessary background information; (3) a "Let's Try This Together" section to facilitate a cooperative effort between the child and his or her parent; (4) a "Can We Do This Another Way?" section to promote a sense of multiple methods of solution; and (5) a "What Do You Think?" section to both provide parents with the opportunity to give feedback about the assignment and allow teachers to assess the initiative.

For this particular interactive homework assignment, paper tangram sets need to be sent home along with it. I suggest printing a tangram set for the parent and a tangram set for the child onto cardstock as you did earlier for the engagement workshop. The parent and their child can then cut the pieces out.

As with all of the ideas shared in this book, the content is presented for you to use directly or amend according to your personal preferences and situations as you craft an interactive homework assignment for your student and parent populations. The content presented below for each of the sections of the interactive homework assignment can flow after each other.

The only section that should appear alone on a separate page is the parent feedback section so that it can be detached and reviewed separately from the actual assignment.

Dear Math Learning Partner,
Here is some homework we can do together using tangrams. I hope you like it. This assignment is due _____.
    Thanks for working with me.
Your Math Learning Partner,

_____
Student's Signature

Take a Look

Remember when we worked with the tangram set? It had seven pieces. Attached to this assignment is a set for each of us. We can cut them out and work with them to complete this homework assignment together.

    Tangram pieces can be different from one another in shape and in area. For example, the square and parallelogram are different shapes, but they have the same area. They have the same area because at our engagement workshop we discovered that these two different shapes are both made of the two small triangles in our tangram set and therefore take up the same amount of space. When shapes take up the same amount of space, they have the same area.

Is there another example of two tangram pieces that are different shapes but have the same area? If so, we'll need to trace the shapes on a separate piece of paper using our tangram pieces. Underneath the shapes, we need to write about what shapes they are and why they have the same area. Let's do that now and then come back and read more.

The small triangle and the medium triangle are the same shape but do not have the same area because they take up different amounts of space. At our engagement workshop, we learned that one is double the size of the other. We can also say that one is half the size of the other.

Is there another example of two tangram pieces that have the same shape but have different areas? If so, we'll need to trace the shapes on our separate piece of paper using our tangram pieces. Underneath the shapes, we'll need to write about what shape they are and why they have different areas. Let's do that now and then come back and read more.

The large triangle and the square are different shapes and have different areas. At our engagement workshop, we discovered how the square can be made with the two small triangles. We also discovered that the large triangle can be made with four small triangles or with the square and two small triangles. It can also be made with the medium triangle and two small triangles or even with the parallelogram and two small triangles.

Is there another example of two tangram pieces that are different shapes and have different areas? If so, we'll need to trace the shapes on our separate piece of paper using our tangram pieces. Underneath our shapes, we'll need to write about what shapes they are and why they have different areas. Let's do that now and then come back and read more.

Let's Try This Together

We can form a Tangram Train by putting tangram pieces together in a row. Each tangram piece represents a car on the train and is different from the one before it in only one way. It can be different from the one before it by shape or area.

Let's each make a Tangram Train. Pick any tangram piece to start with and then add on as many cars (tangram pieces) as you can. Remember, each tangram piece has to be different from the one before it in only one way (by shape or area).

Now let's share and explain our trains to each other. We'll need to double check each other to make sure each of the cars on the train is different from the one before it in only one way. When we are done sharing, we'll need to draw each of our trains onto another separate piece of paper by tracing our tangram pieces and labeling the trains with the name of the person who created each of them. Let's do this now and then come back and read more.

Can We Do This Another Way?

Let's try to make a train that is different from the trains we have already made. Can it be done? If so, we'll need to draw our train on another separate sheet of paper and write about how the pieces are different from one another in only one way.

What Do You Think?

Parent Feedback: Please let my teacher know how we did together with this assignment.

_____ My child did well. The idea of this assignment was understood, and its tasks were successfully completed.

_____ My child needed help. The idea of this assignment seems to be understood, and its tasks seem to be successfully completed.

_____ My child needs instruction. The idea of this assignment was not understood, and its tasks were not successfully completed.

Comments:
_____
_____
_____

Parent's Signature:_____

Classroom Follow-up Techniques

In this section, I discuss ways that you can strengthen the parental involvement sparked by this initiative's efforts through use of group discussion and written reflection in the mathematics classroom. Reflecting on the experience of working with their parents allows students to review the components of the initiative, organize their thoughts, and give insight into what has been successful and what needs to be improved.

To maximize the potential of the interactive homework assignment and foster communication among students about learning at home with their parents, pair the students so that they can share their interactive homework assignments and offer feedback to each other as you circulate the room and give assistance where needed.

Please note that the students should not be paired until you have reviewed their work, detached the parent feedback sheet, and given assistance where needed so that all have accurate work to share.

To initiate the discussions, engage the pairs in describing their two tangram trains (one created by child and one by parent) to each other. They should then compare their trains and discuss how they are alike and different. This reinforces their conceptual understanding and skill with the assignment's task. To close the discussions, have the pairs discuss what was most enjoyable and most challenging about doing the interactive homework assignment with their parent.

Once you notice that your students have adequately discussed their work, have them share their thoughts about the most enjoyable and most challenging aspects of working with their parent in a whole group setting. Through such discussion, students interact as a community of learners who are growing with their parents as partners in mathematical learning.

In addition to group discussions about the interactive homework assignment, I have found it helpful to engage the students in writing letters to their parents concerning the engagement workshop and their learning experiences at home. These letters can in turn be used to guide your efforts in planning future initiatives and be sent home to guide parents' efforts in productively collaborating with their children.

For the younger students whose writing skills may still be developing, drawing a picture that depicts their feelings about doing mathematics with their parent can be helpful in gathering student feedback and guiding your assessment.

Following is a template for student letters for you to use as is or amend according to your individual preferences and situations.

Dear Math Learning Partner,
Thank you for helping me learn math. The most enjoyable part of the engagement workshop was_____
because_____
The most challenging part of the engagement workshop was_____
_____
because_____
The most enjoyable part about learning with you at home was _____
_____
because_____
The most challenging part about learning with you at home was_____
_____
because_____
I'm glad we are math learning partners.
Sincerely,

Interactive Newsletter

I decided on the name "interactive newsletter" because this form of written information stimulates active communication among parents, children, and teachers. The newsletter, titled "Math Notes," serves to review the events of the parental involvement initiative, provide additional support for learning at home with tangrams, share classroom happenings related to tangrams, and open lines of communication with parents to an even greater extent.

Even parents who may not have participated in this initiative may be intrigued by reading the newsletter and in turn choose to involve themselves in future initiatives. The content presented below is for your use while crafting your own interactive newsletters. I suggest placing your content into actual newsletter formats available through software programs such as Microsoft Word, Publisher, and PrintMaster.

Math Notes

Dear Parents,

This interactive newsletter serves to review the events of our parental involvement initiative titled "Using Hands-on Moveable Objects," provides additional support for learning at home with tangrams, and shares classroom happenings related to this topic.

Newsletter contents:

- Workshop Events Revisited
  - Initial Meeting
  - Engagement Workshop
  - Follow-up Session
- Website Information
- Children's Literature Information
- Additional Resources for Home Use
- Classroom Happenings
- Your Comments

Workshop Events Revisited:

Our parental involvement initiative has involved three events (initial meeting, engagement workshop, and follow-up session). The initial meeting informed us about

- changes in mathematics teaching through the years;
- a constructivist foundation to teaching mathematics;
- the value of parental involvement;
- ways that parents can productively collaborate with their children at home to support mathematics education reform efforts;
- the value of manipulatives in the mathematics classroom; and
- specific uses of tangrams in mathematical learning.

Our engagement workshop actively involved us in using tangrams with the children to

- explore existing relationships among the tangram pieces; and
- engage in problem solving that involved visualization, spatial reasoning, and computational skills.

As we interacted with each other in both small and large groups, we communicated about our thinking and shared different methods of solution. Participants left the workshop with an activity and supporting materials to be used at home.

Our follow-up session was filled with reports from parents and their children concerning the home activity distributed at the engagement workshop. We reflected on the experience and grew as a community of mathematical learners.

Website Information:
Use the website www.nctm.org, and search for tangram activities to do with your child that extend our work together and enrich your home learning environment. When searching, type in tangrams into the white search box on the top right-hand corner of the webpage. Click "search" and the activities are listed for you to peruse and engage in together. You can even narrow your search by checking off the grade level(s) that pertain to your child.

Children's Literature Information:
The book titled *Grandfather Tang's Story*, by Ann Tompert, is a resource that allows you to engage in reading with your child as well as using tangrams to explore the many ways that tangrams can be arranged to form various animals. Although this book appears to be focused on younger children, older children can use this book as well as a springboard to crafting their own stories that involve manipulations with tangrams.

They can even create their own word problems by assigning a monetary value to a tangram piece in the same manner we did during the engagement workshop. Total costs of the animals designed with all or some of the tangram pieces can then be calculated.

Additional Resources for Home Use:
To enhance your mathematics home learning environment, the following resources are suggested:

- "Helping Your Child Learn Mathematics" is a free booklet for parents to help strengthen children's mathematical skills and build positive atti-

tudes toward mathematics. Activities are suitable for children in grades Pre-K to 5 and can be found at www.ed.gov/parents/academic/help/math/index.html.
- The website of the National Council of Teachers of Mathematics features *Figure This*, a resource consisting of middle school math challenges for families to explore. This feature also includes a "Family Corner" focusing on parental concerns and questions for all grade levels. This feature can be located at www.figurethis.org
- A book that explores geometric shapes and relationships is *A Cloak for the Dreamer*, by Aileen Friedman. Geometric designs appear throughout the story involving a tailor and his sons who sew pieces of cloth together to make cloaks.

It is a nice resource to follow up on the spatial reasoning that permeated our engagement workshop, home activity, and follow-up session as well as the geometric explorations involved in *Grandfather Tang's Story* mentioned in the Children's Literature section of this newsletter.

Classroom Happenings:
Tangrams continue to be used in your child's mathematics classroom. Your children are reflecting on their collaboration with you concerning their mathematical learning in the following ways:

- group discussions concerning the interactive homework assignment
- written reflections concerning their collaboration with you

Your Comments:
To open our lines of communication even further, please share your feedback concerning any part of the parental involvement initiative. Feel free to suggest topics for future initiatives and to express concerns that can be addressed to help you nurture your home learning environment.
Please share your comments below and return to school by_____.

Sincerely,

SUGGESTIONS FOR FURTHER READING

Adams, J. (2002). *Grades k–8: The super source tangrams*. Vernon Hills, IL: ETA/Cuisenaire.

Del Grande, J. (1993). *Geometry and spatial sense: Curriculum evaluation standards for school mathematics addenda series, grades K–6*. Reston, VA: NCTM.

Geddes, D. (1992). *Geometry in the middle grades: Curriculum evaluation standards for school mathematics addenda series, grades 5–8*. Reston, VA: NCTM.

Rectanus, C. (1994). *Math by all means: Geometry grade 2*. Sausalito, CA: Math Solutions Publications.

Rectanus, C. (1994). *Math by all means: Geometry grade 3*. Sausalito, CA: Math Solutions Publications.

Winter, M., Lappan, G., Phillips, E., & Fitzgerald, W. (1986). *Middle grades mathematics project: Spatial visualization*. Menlo Park, CA: Addison Wesley.

REFERENCES

Burns, M. (1996). How to make the most of manipulatives. *Instructor*, 105(7), 45–51.

De La Cruz, Y. (1999). Reversing the trend: Latino families in real partnerships with schools. *Teaching Children Mathematics*, 5(5), 296–300.

Epstein, J. L. (2001). Practical applications: Linking family and community involvement to student learning. In *School, family, and community partnerships: Preparing educators and improving schools*. Boulder, CO: Westview Press, 507–49.

Epstein, J. L., & Van Voorhis, F. L. (2002). Implementing teachers involve parents in schoolwork (TIPS). In *School, family, and community partnerships: Your handbook for action*. Thousand Oaks, CA: Corwin Press, 289–320.

ETA/Cuisenaire (n.d.). Tangrams make cents. In *The super source*. Vernon Hills, IL: ETA/Cuisenaire, 74–77.

Fuys, D., & Tishler, R. (1979). *Teaching mathematics in the elementary school*. New York: HarperCollins.

Guastello, E. F. (2004). A village of learners. *Educational Leadership*, 61(8), 79–83.

Mistretta, R. M. (2004). Parental issues and perspectives concerning mathematics education at elementary and middle school settings. *Action in Teacher Education*, 26(2), 69–76.

National Council of Teachers of Mathematics (NCTM). (2000). *Principles and standards for school mathematics*. Reston, VA: Author.

Chapter Four

Interacting with Web-Based Resources

Linking technology with curriculum has caused significant changes in teaching and learning. Wright (1999) reports higher student achievement, self-concept, attitude, and teacher-student interaction as a result of interactive learning made possible via technology. Neiss (2001) reports the National Council of Teachers of Mathematics (NCTM) as pinpointing technology as an essential component of the Pre-K to 12 mathematics learning environment due to its ability to influence the mathematics that is taught and in turn enhance students' learning.

Kerrigan (2002) has found the capabilities of mathematics websites to include promoting higher-order thinking skills, developing and maintaining computational skills, introducing children to collecting and analyzing data, facilitating algebraic and geometric thinking, and showing the role of mathematics in an interdisciplinary setting.

To increase awareness and understanding of this mathematics learning tool among parent populations, this chapter describes a parental involvement initiative designed to empower parents' awareness of and confidence in using web-based resources to help their children develop understanding of mathematical concepts, computational fluency, and problem-solving ability.

The structure of this chapter reflects that of chapter 3. An invitation to parents along with explanations and related materials concerning the initiative's initial meeting, engagement workshop, follow-up session, and methods of maintaining the home-school mathematics connection are its components.

Technology is a major feature of the world in which our current students live. Its proper use is essential both in school and at home. By building parents' understanding of web-based resources, this chapter's described initiative cultivates another venue for productive parent-child collaboration in mathematical learning.

ENCOURAGE YOUR PARENTS TO COME

As discussed in previous chapters, inviting parents into a community of learners builds meaningful partnerships. Following is the text for the letter of intent and inquiry that informs parents of the initiative's intent, requests their commitment to participate, and inquires about times that would best suit their busy schedules.

It is presented at this point for you to use when crafting your own letter of intent and inquiry or amend according to your specific situations. As described in chapter 3, personalize the invitation by having students design their own cover for it. Also included are a response form for you to use when gathering information about appropriate meeting times and a letter of announcement for you to use when beginning this initiative.

Letter of Intent and Inquiry

Dear Parents,
Children need to find mathematics experiences interesting and meaningful to reach their full mathematics potential. Using web-based resources has proven to be successful in developing students' ability to understand mathematics concepts, master skills, and problem solve. Your involvement with such a practice plays a vital role in supporting your child's mathematics education.

To empower you as partners with this aspect of your child's education, a parental involvement initiative titled "Interacting with Web-Based Resources" has been designed for you and your child. Instruction on appropriately choosing and using websites for mathematical learning at home permeate the initiative. Criteria to evaluate websites are shared as well as specific websites conducive to sound mathematical learning.

This initiative involves an initial meeting (1 hour), an engagement workshop (1 1/2 hours), home activities, a follow-up session (1 hour), and additional tasks to help maintain connections between the mathematics classroom and your home. The initiative *informs* you about the importance of parental involvement and the rationale behind specific teaching methods used in today's mathematics classrooms.

The initiative *engages* you and your child in activities involving the use of web-based resources in a workshop setting that extends into your home. Time for *reflection* on your learning experiences is facilitated as well between you and your child, among the parent community, and among the student community.

Please be aware that you are expected to attend the initial meeting and engagement workshop, engage in home activities with your child, attend the follow-up session, and maintain connections between the mathematics classroom and your home. Three meetings at school are required, where one is attended only by parents and two are attended by both parent and child.

If you wish to commit to this parental involvement initiative, please return the attached form. If you have more than one child, please arrange for each child to be represented by one family member.

We want to schedule this initiative in consideration of the busy schedules of all those involved. Therefore, please indicate on the attached form the time that would best suit you.

I look forward to working with you.

Sincerely,

The text of the response form that is mentioned in the above letter of intent and inquiry is presented below. As described in chapter 3, it fosters a sense of commitment, serves to investigate the most appropriate meeting times for your parent population, allows parents to indicate reasons they may not be able to participate, and serves as an organizational tool when keeping accurate records of attendance throughout the initiative.

Following the text of the response form is text for a letter of announcement. Spaces are provided for specific dates for the initial meeting, engagement workshop, and follow-up session. As mentioned previously, the text is shared with you for your direct use in creating your own response form and letter of announcement or as a framework to amend according to your personal preferences.

Response Form

Please return this form by _____.
We (commit, cannot commit) to the parental involvement initiative titled "Interacting with Web-Based Resources."
If you cannot commit to this initiative, please explain why on the back of this form so we can try to partner with you in another way.
Name of Parent(s) and Other Family Members Committed to the Initiative (Each child must be accompanied by one family member.)
_____
_____
_____

Name of Child(ren) (with grade level) Committed to the Initiative (Indicate the family member above to partner with each child indicated below.)
_____
_____
_____
_____

Parent(s) Signature(s): _____
_____

Student(s) Signature(s): _____
_____

Circle the most appropriate time for your family to commit to this initiative.
Weeknight beginning at 7:00 P.M. (Specify night) _____
Weeknight beginning at 8:00 P.M. (Specify night) _____
Saturday morning beginning at 10:00 A.M. _____
Saturday afternoon beginning at 1:00 P.M. _____
Sunday morning beginning at 10:00 A.M. _____
Sunday afternoon beginning at 1:00 P.M. _____

Letter of Announcement

Dear Parents/Family Members,

To empower you as partners in your child's learning of mathematics, the parental involvement initiative titled "Interacting with Web-Based Resources" has been planned for you and your child. Your feedback concerning the most appropriate days and times for this event have been carefully considered, and the following dates and times reflect common responses.

Initial Meeting _____
(1 hour session for parents only)
Engagement Workshop _____
(1 1/2 hour session for parents and children)
Follow-up Session _____
(1 hour session for parents and children)

As noted in previous correspondence, the initiative *informs* you about the importance of parental involvement and the rationale behind specific teaching methods used in today's mathematics classrooms. The initiative *engages* you and your child in activities involving the use of web-based resources in a workshop setting that extends into your home. Time for *reflection* on your learning experiences is facilitated between you and your child, among the parent community, and among the student community as well.

Please be aware that you have committed to this initiative and are expected to attend the initial meeting and engagement workshop, engage in a home activity with your child, attend the follow-up session, and maintain connections between the mathematics classroom and the home. All meetings start exactly on time, so please be punctual.

I look forward to working with you.

Sincerely,

INITIAL MEETING

As stated in chapter 3, the initial meeting should be attended by parents only due to its content. You should begin the meeting by presenting the initiative's goals and components using the information described in chapter 2 (Key Points to Share with Parents). This introduction allows parents to become knowledgeable of the initiative's structure and features that serve to inform, engage, promote reflection, and maintain connections between the mathematics classroom and the home.

After this introduction, a discussion can follow with the use of the organized information discussed in chapter 1 (Key Points to Share with Parents) about the changes in mathematics teaching through the years, a constructivist foundation to teaching mathematics, the value of parental involvement, and the ways that parents can productively partner with their child at home to support mathematics education reform efforts. The discussion can then turn to the topic of using web-based resources in mathematical learning.

Please note that if your parent populations have participated in the initiative discussed in chapter 3, the introductory material from chapters 1 and 2 may be repetitious. If this is the case, just reminding parents of information covered in the prior initiative and discussing the following information may be more appropriate. I leave that decision up to you and your individual settings.

Whatever direction you take, the next step is to inform parents that web-based resources positively affect mathematical learning at all grade levels, and it is important that they partner with teachers in developing students' appreciation of technology as a learning tool (Bitter & Pierson, 2005). Helpful information is organized below to facilitate your discussion about what the Internet actually is, the rationale for using web-based resources in education (Bitter & Pierson, 2005), and the criteria parents should use when evaluating websites (Roblyer, 2003).

As with the information taken from chapters 1 and 2, this content should be used to create transparencies, PowerPoint presentations, and hand-outs as well so that parents have materials to refer to both during this initiative's events as well as at home.

- What is the Internet?
    - Represents a worldwide collection of computer networks, connected by special phone lines, satellites, microwave relays, fiber optics, and sophisticated software
    - Allows a global community to form with the common vision of sharing knowledge and information.

- Why use web-based resources?
  - Motivation
    - Gains learner attention
    - Engages learner in productive work
    - Increases perceptions of control (can cover material as slow or as fast as needed)
  - Support
    - Facilitates cooperative learning environments
    - Provides access to various knowledge bases
    - Builds conceptual understandings and procedural skills
    - Develops problem-solving and higher-level thinking skills
- What criteria should one use when evaluating web-based resources?
  - Content relevant to your needs
  - Information reliable, up to date, and accurate
  - Ease of navigation
  - Appropriate reading level for user
  - Site provides useful links to other sites

Let the parents know they'll become engaged in accessing websites with their children during the engagement workshop and at home that build conceptual understanding, computational skills, and problem-solving ability. Inform them that they'll concretely experience these web-based resources as learning tools for mathematics.

As in the initiative described in chapter 3, allow a question-and-answer session to follow this discussion so that parents can contribute feedback and clarify any existing concerns. If questions from a small number of parents require in-depth responses and do not pertain to all those present, speak with these select parents after the workshop to avoid detaining others.

Before closing, remind parents that the engagement workshop involving both them and their children takes place two weeks later at the same time. As previously stated in chapter 3, I've found sending a reminder home to parents a day or two before this event to be helpful. A two-week interval between the initial meeting and engagement workshop gives parents enough time to review the material presented at the initial meeting and remain vibrant about the initiative's intentions.

ENGAGEMENT WORKSHOP

As in the initiative described in chapter 3, you are at a point now where you have given your parents the background information they need to understand the rationale behind current mathematics teaching methods, specifically the use of web-based resources, and the value of their involvement in their child's mathematics education.

The engagement workshop can in turn take on a different environment that is quite active and insightful for parents and their children, and even you, as you explore and communicate about mathematics with the use of web-based resources.

Obviously, this workshop needs to take place in a room with sufficient computers, Internet access, and a projector for your use in demonstrating the features of particular websites. Begin the engagement workshop by giving an overview to parents and children concerning the websites stated below. These websites have proven to be quite beneficial for me in getting children and parents to work together on mathematical tasks. They are as follows:

- National Council of Teachers of Mathematics (www.nctm.org)
- Funbrain (www.funbrain.com)
- Math Cats (www.mathcats.com)

These websites may even be new for you. If so, first explore them thoroughly yourself in the same manner I describe below for you to follow with the parents and children. Of course, feel free to venture out into the other areas of the sites. There are many activities that you can incorporate into this initiative and also use in your everyday teaching of mathematics.

When viewing the website for the NCTM, point out specific features in the families section such as "Figure This," a section that offers activities for middle school students. Specifically, view the "Challenge Index" and the "Family Corner" so that families can get acquainted with the structure of this section.

Showcase how the "Illuminations" section not only offers online activities that help energize mathematics in the classroom and at home but also provides information concerning the *Principles and Standards for School Mathematics* (NCTM, 2000) plus lists numerous Web links to exemplary online resources.

Next, demonstrate how Fun Brain can aid in computational skill building with its twenty-five games that involve addition, subtraction, multiplication, and division. These games are motivational and track the child's progress as they proceed through them. Point out how the child's grade level can be indicated so that the games pertain to specific grade level computational objectives.

Now present Math Cats with its components that consist of thinking games, projects, crafts, an art gallery, interdisciplinary connections, and problem-solving situations authored by children of all ages across the United States. This website was designed to promote open-ended and playful explorations of important mathematical concepts. There is also a special section, titled "Older Cats," that serves as a resource for parents seeking ideas on how to help their children with various mathematical curriculum topics.

Instruct the group that at this point they are given three fifteen-minute tasks to complete (one from each website) according to their grade level group, namely Pre-K to 2, 3 to 5, and 6 to 8. These grade level tasks and navigating instructions are presented below. Please note that the task involving the National Council of Teachers of Mathematics' website involves the use of tangrams and is a very nice reinforcement of the initiative discussed in chapter 3 for all grade levels.

I suggest creating hand-outs to be used during this part of the engagement workshop that include the tasks that pertain to the grade level(s) you are working with.

Also, before allowing families to explore on their own, it is wise to navigate through each website with them using the instructions presented below in order to get them acquainted with the areas of the websites.

- Mathematics Tasks for Grades Pre-K to 2
  - National Council of Teachers of Mathematics
    - Go to www.nctm.org.
    - Click on Principles and Standards for School Mathematics.
    - Scroll down the left side of the page, and click on E-examples.
    - Explore the E-example titled "Developing Geometry Understandings—Tangram Puzzles."
    - Try to organize the tangram pieces into one of the given outlines.
    - See if there is more than one way to solve the puzzle.

- Funbrain
  - Go to www.funbrain.com.
  - Scroll down to the statement "Find a game by grade."
  - For grades Pre-K and K, click on grade K and choose "Oddball."
  - For grades 1 and 2, click on grade 1 or grade 2 and choose "Math-Car Racing."
  - Play the games and have fun.
- Math Cats
  - Go to www.mathcats.com.
  - Click on the sign saying Math Cats to enter the website.
  - Click on "Math Cats Explore."
  - Scroll down.
  - For grades Pre-K and K, click on "Tessellation Town on Tile Island" and have fun making tessellations. For grades 1 and 2, click on "Math Story Problems" and solve as many of the "Kitten" and "House Cat" problems as you can. Even create your own story to be published on the website. To do this, scroll down the story page to where it says "Submit your own math story problem here."

- Mathematics Tasks for Grades 3 to 5
  - National Council of Teachers of Mathematics
    - Go to www.nctm.org.
    - Click on "Principles and Standards for School Mathematics."
    - Click on E-examples.
    - Explore the E-example titled "Developing Geometry Understandings—Tangram Challenges."
    - Try to find solutions to as many challenge questions as you can.
  - Funbrain
    - Go to www.funbrain.com.
    - Scroll down to the statement "Find a game by grade."
    - For grade 3, choose "Measure It."
    - For grades 4 and 5, choose one of the several math games listed.
    - Play the games and have fun.
  - Math Cats
    - Go to www.mathcats.com.
    - Click on the sign saying Math Cats to enter the website.
    - Click on "Math Cats Explore."

❏ Scroll down and click on "Math Story Problems." Solve as many of the "Alleycat" and "Leopard" problems as you can, and even create your own story to be published on the website. To do this, scroll down the story page to where it says "Submit your own math story problem here."
- Mathematics Tasks for Grades 6 to 8
  - National Council of Teachers of Mathematics
    ❏ Go to www.nctm.org.
    ❏ Click on "Principles and Standards for School Mathematics."
    ❏ Click on E-examples.
    ❏ Explore the E-example titled "Developing Geometry Understandings—Tangram Challenges."
    ❏ Try to find solutions for as many challenge questions as you can.
  - Funbrain
    ❏ Go to www.funbrain.com.
    ❏ Scroll down to the statement "Find a game by grade."
    ❏ Choose one of the several math games listed.
    ❏ Play the games and have fun.
  - Math Cats
    ❏ Go to www.mathcats.com.
    ❏ Click on the sign saying Math Cats to enter the website.
    ❏ Click on "Math Cats Explore."
    ❏ Scroll down.
    ❏ Click on "Math Story Problems." Solve as many of the "Leopard" and "Tiger" problems as you can, and even create your own story to be published on the website. To do this, scroll down the story page to where it says "Submit your own math story problem here."

As in chapter 3, place emphasis on communication. Advise parents not to do all the telling. Rather, encourage them to communicate and explore their children's thought processes. Also as stated in chapter 3, it is best to take a break after the tasks have been completed. Allow parents and children to stretch and partake in some light refreshments that you provide.

The next step is to facilitate discussion time (approximately ten minutes) for the group to reflect on their experiences using web-based resources. As in the initiative discussed in chapter 3, divide the participants into smaller groups of two to three families and ask them to discuss the reflection questions presented below. Have someone record their responses and designate another to later share the responses with the entire group.

- What did you find most interesting about each of the tasks that involved the three websites: www.nctm.org, www.funbrain.com, and www.mathcats.com?
- Was this experience challenging? If so, how?
- Was this experience fun? If so, how?
- How would you use the websites at home?

Once you notice that the majority of smaller groups have answered the reflection questions, as in the initiative discussed in chapter 3, bring the families back together as an entire group to share their thoughts and listen to the experiences of others (approximately fifteen minutes). Request that the person designated to present the responses come and lead the conversation about their group's experiences. Seats should be arranged so that the other members of the group can contribute to the discussion as well.

Parents have often voiced their surprise at the bountiful web-based resources available for mathematical learning. They valued the guidance on how to choose appropriate websites and found the opportunities to work with their children on mathematical tasks using web-based resources "exciting and supportive of their needs."

The children were equally impressed with the websites. They expressed their contentment with "being able to control how fast or how slow they moved through the activities." They often commented, "There's a place for everything and everyone on these math websites!"

To both reinforce and extend this workshop experience, the stage now needs to be set for learning at home. A cover letter that is appropriate for everyone and grade-level home activities that reinforce and extend the engagement workshop experience are provided for you at this point. You may use the content as is or amend it according to your specific situations as you develop your materials for a home learning experience.

In closing the engagement workshop, distribute and review these materials with everyone so that the parents and children are clear about what needs to be done as they engage in doing mathematics together at home using web-based resources.

Specific website addresses have been included with the activities since you are not there to guide the parents through the websites as you did during the engagement workshop. In some instances, I have elaborated for the parents on what to do in the activities since directions were minimal on the website.

Instruct parents to collaborate with their children in the same manner they did during the engagement workshop. Remind them to not do all of the telling but rather communicate and explore together. Most importantly, emphasize the importance of just having fun at home with mathematics.

After the activities have been explored for each of the websites, parents are to complete a reflection paper concerning the experience. This reflection paper is included for your use on the following pages. Inform the parents that this reflection paper and their responses must be brought to the follow-up session so that discussion as a community of mathematics learners using web-based resources can be facilitated.

Cover Letter

Dear Parents and Students,

Presented to you in this letter are activities for you to do at home that both reinforce and extend our learning from the engagement workshop. The activities relate to the three websites we explored together during the engagement workshop. Remember to again share your thinking as you progress through the activities. Parents, remember that if you do all of the telling this is not a productive experience.

After you complete each of the activities, complete the attached reflection paper as well. Our follow-up session takes place on _____ at _____ so we can discuss your experiences. Please bring the reflection papers with you. This session (approximately one hour) starts exactly on time, so please be punctual.

I look forward to discussing with you your reactions to the activities. As you work together, keep in mind the importance of exploring mathematics together, sharing your thinking, and just having fun.

Sincerely,

Activities for Grades Pre-K to 2

Explore the following three websites:

- www.nctm.org
- www.funbrain.com
- www.mathcats.com

For each website, do the indicated activity. In case you have trouble finding the activities, the web address for each specific activity has also been indicated for you.

After an activity is completed for each of the three websites, complete the question related to the activity's website on the attached reflection paper.

- www.nctm.org
  - As in the engagement workshop, go to www.nctm.org, click on "Principles and Standards for School Mathematics," scroll down the left side of the page and click on E-examples. Click on Pre-K to 2, and try one of the following activities:
    - Creating, Describing, and Analyzing Patterns
      http://standards.nctm.org/document/eexamples/chap4/4.1/index.htm
    - Investigating the Concept of Triangles
      http://standards.nctm.org/document/eexamples/chap4/4.2/index.htm
    - Learning Geometry and Measurement Concepts
      http://standards.nctm.org/document/eexamples/chap4/4.3/index.htm
    - Learning About Number Relationships
      http://standards.nctm.org/document/eexamples/chap4/4.5/index.htm
    - Developing Estimation Strategies
      http://standards.nctm.org/document/eexamples/chap4/4.6/index.htm
- www.funbrain.com
  - As in the engagement workshop, go to www.funbrain.com. Click on "All Games" under "Classic Funbrain." Scroll down to the statement "Find a game by grade" and play:
    - Measure It
      www.funbrain.com/measure/index.html
- www.mathcats.com
  - As in the engagement workshop, go to www.mathcats.com, click on the "MathCats" sign, and then on "Math Cats Explore." Try one of the following activities:

❑ Number Stories
www.mathcats.com/explore/numberstories.html
Have your child read through the stories and look at the pictures that depict the stories. Have them draw their own picture for the story. Have them write/tell a math story and draw a picture for it.
❑ Polygon Playground
www.mathcats.com/explore/polygons.html
Here you and your child can use various polygons to build a playground. Encourage your child to try to fit the polygons together where possible and notice the differences among them.

Activities for Grades 3 to 5

With your child, explore the following three websites:

- www.nctm.org
- www.funbrain.com
- www.mathcats.com

For each website, do the indicated activity. In case you have trouble finding the activities, the web address for each specific activity has also been indicated for you.

After an activity is completed for each of the three websites, complete the question related to the activity's website on the attached reflection paper.

- www.nctm.org
  - As in the engagement workshop, go to www.nctm.org, click on "Principles and Standards for School Mathematics," scroll down the left side of the page, and click on E-examples. Click on 3 to 5, and try one of the following activities:
    ❑ Communicating About Mathematics Using Games
    http://standards.nctm.org/document/eexamples/chap5/5.1/index.htm
    ❑ Understanding Distance, Speed, and Time
    http://standards.nctm.org/document/eexamples/chap5/5.2/index.htm
    ❑ Exploring Properties of Rectangles
    http://standards.nctm.org/document/eexamples/chap5/5.3/index.htm
    ❑ Accessing and Investigating Data
    http://standards.nctm.org/document/eexamples/chap5/5.4/index.htm

- ❏ Collecting, Representing, and Interpreting Data
  http://standards.nctm.org/document/eexamples/chap5/5.5/index.htm
- www.funbrain.com
  - As in the engagement workshop, go to www.funbrain.com. Click on "All Games" under "Classic Funbrain." Scroll down to the statement "Find a game by grade," and play one or more of the following games. Adjust the level of difficulty (Easy, Medium, Hard, or Super-brain) to suit your needs.
    - ❏ Math Baseball
      www.funbrain.com/math/index.html
    - ❏ Tic Tac Toe Squares
      www.funbrain.com/tictactoe/index.html
    - ❏ Power Football
      www.funbrain.com/football/index.html
    - ❏ Change Maker
      www.funbrain.com/cashreg/index.html
    - ❏ Fresh Baked Fractions
      www.funbrain.com/fract/index.html
- www.mathcats.com
  - As in the engagement workshop, go to www.mathcats.com, click on the "MathCats" sign, and then on "Math Cats Explore." Try one of the following activities:
    - ❏ Crossing the River
      www.mathcats.com/explore/river/crossing.html
    - ❏ Old Egyptian Math Cats Fractions
      www.mathcats.com/explore/oldegyptianfractions.html
    - ❏ Math-Loving Animals in the News
      www.mathcats.com/explore/animalsinthenews.html
    - ❏ Interactive Multiplication Table
      www.mathcats.com/explore/multiplicationtable.html

Activities for Grades 6 to 8

With your child, explore the following three websites:

- www.nctm.org
- www.funbrain.com
- www.mathcats.com

Interacting with Web-Based Resources 87

For each website, do the indicated activity. In case you have trouble finding the activities, the web address for each specific activity has also been indicated for you.

After an activity is completed for each of the three websites, complete the question related to the activity's website on the attached reflection paper.

- www.nctm.org
  - As in the engagement workshop, go to www.nctm.org, click on "Principles and Standards for School Mathematics," scroll down the left side of the page, and click on E-examples. Click on 6 to 8, and try one of the following activities:
    - Learning About Multiplication
      http://standards.nctm.org/document/eexamples/chap6/6.1/index.htm
    - Learning About Rate of Change
      http://standards.nctm.org/document/eexamples/chap6/6.2/index.htm
    - Learning About Length, Perimeter, Area, and Volume
      http://standards.nctm.org/document/eexamples/chap6/6.3/index.htm
    - Understanding Congruence, Similarity, and Symmetry
      http://standards.nctm.org/document/eexamples/chap6/6.3/index.htm
    - Understanding the Pythagorean Relationship
      http://standards.nctm.org/document/eexamples/chap6/6.4/index.htm
    - Comparing Properties of the Mean and the Median
      http://standards.nctm.org/document/eexamples/chap6/6.4/index.htm
- www.funbrain.com
  - As in the engagement workshop, go to www.funbrain.com. Click on "All Games" under "Classic Funbrain." Scroll down to the statement "Find a game by grade," and play one or more of the following games. Adjust the level of difficulty (Easy, Medium, Hard, or Superbrain) to suit your needs.
    - Shape Surveyor
      www.funbrain.com/poly/index.html
    - Soccer Shootout
      www.funbrain.com/fractop/index.html
    - Guess the Number Plus
      www.funbrain.com/guess2/index.html
    - Operation Order
      www.funbrain.com/algebra/index.html

- www.mathcats.com
  - As in the engagement workshop, go to www.mathcats.com, click on the "MathCats" sign, and then on "Math Cats Explore." Go to "Math Story Problems," and solve more of the leopard and tiger problems. Write your own story problem and submit it to the website. www.mathcats.com/storyproblems.html

Reflection Paper

Answer the following questions as you use each of the web-based resources. Bring your written responses with you to our follow-up session.

- In what ways was the website www.nctm.org most helpful?
- In what ways was the website www.mathcats.com most helpful?
- In what ways was the website www.funbrain.com most helpful?
- What did you find most interesting about using the websites?
- How has this experience impacted your family?

FOLLOW-UP SESSION

As stated in chapter 3, allow two weeks before holding the follow-up session so that parents have enough time to complete the activities with their child in a manner that is not rushed. I've found that sending a reminder letter home a week before helps alert parents to this upcoming event. The main goal of this session is to share families' experiences with the web-based resources and reflect as a community of learners.

Placing families into the same smaller groups they were in during the engagement workshop is beneficial since there is already an established sense of familiarity among the group members. Participants are now at a stage where they can continue where they left off and share their progress.

Parents and children should share their responses to the reflection paper questions (approximately twenty minutes). Designate two people as you did during the engagement workshop to record responses and report on them. When you notice that all has been accomplished within the groups, bring the entire group together for a discussion, and record their responses either on chart paper or a transparency.

As stated in the follow-up session of the initiative discussed in chapter 3, such reflection empowers parents to support each other as a community of mathematics learners. During my implementation of this initiative, comments included "I didn't realize the wealth of resources available on the web," and "My child's world involves technology, and I now can connect his enthusiasm for the tool with quality learning."

Take the remaining time to both discuss the ways that the home-school connection is maintained and share this entire initiative's connections to the technology component of current mathematics curricula. Explanations of the interactive homework assignment, classroom follow-up techniques, and the interactive newsletter that serve to maintain the home-school connection can be found at the end of the follow-up section of chapter 3. As stated there, the content can be used for transparencies, PowerPoint presentations, and hand-outs.

Highlight that technology enhances mathematics learning, and stress with parents that they need to partner with teachers so that technology use at home can support effective mathematics teaching in the classroom. Also point out that the web-based resources used throughout the initiative address children's conceptual understanding of mathematical ideas, computational skills, and problem-solving abilities.

MAINTAINING THE CONNECTION

To ensure continuity of the strides you have made during this initiative, your students and their parents need to remain connected. In this section, I share, as I did in chapter 3, ways to accomplish this through an interactive homework assignment, classroom follow-up techniques, and an interactive newsletter designed to keep the home-school connection alive.

Interactive Homework

As stated in chapter 3, interactive homework engages both the child and their parent in a mathematics task where input from both parties creates the final product. It gives parents the opportunity to communicate not only with their child about their reasoning but also with their child's teacher about the progress made with the assignment and any existing concerns.

The interactive homework assignment on the following pages engages parents and children in using the web-based resources that they did not yet use during this initiative.

As in chapter 3, the format consists of (1) a "Dear Math Learning Partner" section where the parent is invited to work with their child on a mathematics task; (2) a "Take a Look" section to give necessary background information; (3) a "Let's Try This Together" section to facilitate a cooperative effort between the child and their parent; (4) a "Can We Do This Another Way?" section to promote deeper interactive experiences with web-based resources; and (5) a "What Do You Think?" section to provide parents with the opportunity to give feedback about the assignment and allow teachers to assess this component of the initiative.

As stated previously, the content is presented for you to use directly or amend according to your personal preferences and situations as you craft an interactive homework assignment for your student and parent populations. The content presented below for each of the sections of the interactive homework assignment can flow after each other. The only section that should appear alone on a separate page is the parent feedback section so that it can be detached and reviewed separately from the actual assignment.

Dear Math Learning Partner,
Here is some homework we can do together using web-based resources. I hope you like it. This assignment is due _____.
    Thanks for working with me.

                          Your Math Learning Partner,

                          Student's Signature

Take a Look

Remember when we worked with web-based resources from the National Council of Teachers of Mathematics' website (www.nctm.org)? They are listed below for different grade levels. Let's take a look at what each of them asks us to do. We don't have to look at the ones we have already looked at.

- Grades Pre-K to 2
  - Creating, Describing, and Analyzing Patterns
    http://standards.nctm.org/document/eexamples/chap4/4.1/index.htm
  - Investigating the Concept of Triangles
    http://standards.nctm.org/document/eexamples/chap4/4.2/index.htm
  - Learning Geometry and Measurement Concepts
    http://standards.nctm.org/document/eexamples/chap4/4.3/index.htm
  - Learning About Number Relationships
    http://standards.nctm.org/document/eexamples/chap4/4.5/index.htm
  - Developing Estimation Strategies
    http://standards.nctm.org/document/eexamples/chap4/4.6/index.htm
- Grades 3 to 5
  - Communicating About Mathematics Using Games
    http://standards.nctm.org/document/eexamples/chap5/5.1/index.htm
  - Understanding Distance, Speed, and Time
    http://standards.nctm.org/document/eexamples/chap5/5.2/index.htm
  - Exploring Properties of Rectangles
    http://standards.nctm.org/document/eexamples/chap5/5.3/index.htm
  - Accessing and Investigating Data
    http://standards.nctm.org/document/eexamples/chap5/5.4/index.htm
  - Collecting, Representing, and Interpreting Data
    http://standards.nctm.org/document/eexamples/chap5/5.5/index.htm
- Grades 6 to 8
  - Learning About Multiplication
    http://standards.nctm.org/document/eexamples/chap6/6.1/index.htm
  - Learning About Rate of Change
    http://standards.nctm.org/document/eexamples/chap6/6.2/index.htm
  - Learning About Length, Perimeter, Area, and Volume
    http://standards.nctm.org/document/eexamples/chap6/6.3/index.htm
  - Understanding Congruence, Similarity, and Symmetry
    http://standards.nctm.org/document/eexamples/chap6/6.3/index.htm
  - Understanding the Pythagorean Relationship
    http://standards.nctm.org/document/eexamples/chap6/6.4/index.htm
  - Comparing Properties of the Mean and the Median
    http://standards.nctm.org/document/eexamples/chap6/6.4/index.htm

Let's Try This Together

Let's try an activity from the previous list that we haven't done yet. How is it different from the activity we already did together? Let's write about what we just talked about on a separate sheet of paper.

Can We Do This Another Way?

Together let's try to do the activity we just did together another way. Can it be done? If so, what changes did we make? If another way isn't possible, let's do a different activity together and explain what we did on our separate sheet of paper.

What Do You Think?

Parent Feedback: Please let my teacher know how we did together with this assignment.

_____ My child did well. The idea of this assignment was understood, and its tasks were successfully completed.

_____ My child needed help. The idea of this assignment seems to be understood, and its tasks seem to be successfully completed.

_____ My child needs instruction. The idea of this assignment was not understood, and its tasks were not successfully completed.

Comments:
_____
_____
_____

Parent's Signature:
_____

Classroom Follow-up Techniques

Reflecting on the experience of working with their parents allows students to review the components of the initiative, organize their thoughts, and give insight into what has been successful and what needs to be improved. In this section, as in chapter 3, I discuss ways that you can strengthen the parental involvement sparked by this initiative's efforts through use of group discussion and written reflection. If computers with Internet access are not available in your classroom, then you and your students should use the computer lab.

In this setting, pair the students so that they can share their responses on the interactive homework assignments and offer feedback to each other. As stated in chapter 3, this fosters communication among students about learning at home with their parents. Also mentioned previously, the students should not be paired until you have reviewed their responses, detached the parent feedback sheet, and given assistance where needed so that all have valuable feedback to share.

To initiate the discussions, engage the pairs in responding to each other about the following items:

- What activities on the Internet did you do with your parent(s)?
- Show each other what you did.
- What was most enjoyable about doing this assignment with your parent?
- What was most challenging about doing this assignment with your parent?

Once you notice that your students have adequately discussed the questions, have them share responses in a whole group setting to nurture interaction as a community of learners who are growing with their parents as partners in mathematics learning.

In addition to group discussions, I have found it helpful to engage the students in writing letters to their parents. These letters can in turn be sent home to help nurture home learning environments and guide your efforts in planning future initiatives. The format for these letters is included in chapter 3 in the section concerning classroom follow-up techniques. For the younger students whose writing skills may still be developing, have these students draw a picture that depicts their feelings about doing mathematics with their parent.

Interactive Newsletter

I decided on the name "interactive newsletter" because, as mentioned in chapter 3, this form of communication not only informs but also stimulates interaction among parents, children, and teachers. The interactive newsletter, titled "Math Notes," presented at this point serves to review the events of the parental involvement initiative, provide additional support for learning at home with web-based resources, share classroom happenings related to web-based resources, and open lines of communication with parents to an even greater extent.

Even parents who may not have participated in this initiative may be intrigued by reading the newsletter and in turn choose to participate in future initiatives. The content presented below is for your use while crafting your own interactive newsletters. I suggest placing your content into actual newsletter formats available through software programs such as Microsoft Word, Publisher, and PrintMaster.

Math Notes

Dear Parents,
This interactive newsletter serves to review the events of our parental involvement initiative titled "Interacting with Web-Based Resources," provides additional support for learning at home with instructional technology (both websites and software), and shares classroom happenings related to this topic.

Newsletter contents:

- Workshop Events Revisited
    - Initial Meeting
    - Engagement Workshop
    - Follow-up Session
- Website Information
- Software Information
- Additional Resources for Home Use
- Classroom Happenings
- Your Comments

Workshop Events Revisited:
Our parental involvement initiative has involved three events (initial meeting, engagement workshop, and follow-up session). The initial meeting informed us about the following:

- changes in mathematics teaching through the years;
- a constructivist foundation to teaching mathematics;
- the value of parental involvement;
- ways that parents can productively collaborate with their children at home to support mathematics education reform efforts;
- the instructional value and specific uses of web-based resources in mathematical learning.

Our engagement workshop actively involved us in using web-based resources with the children. We engaged in tasks that involved conceptual understandings of mathematical ideas, computational skills, and problem solving. As we interacted with each other in both small and large groups, we shared our findings and experiences. Participants left the workshop with tasks to do at home involving web-based resources.

Our follow-up session was filled with reports from parents and their children concerning the web-based tasks distributed at the engagement workshop. We reflected on the experience and grew as a community of mathematical learners.

Website Information:

Following are some more web-based resources you can access with your child that extend our work together and enrich your home learning environment. These web-based resources can be used for mathematics as well as other subjects.

- www.geom.uiuc.edu/
- www.awesomelibrary.com/
- www.enchantedlearning.com/
- http://nlvm.usu.edu/en/nav/vlibrary.html

Software Information:

The following software titles are useful in learning mathematics and provide enjoyable settings that connect mathematics to our world.

- Grades Pre-K to 2
    - "Winnie the Pooh Math"
    - "Millie's Math House"

- Grades 3 to 5
  - "Hot Dog Stand"
  - "Math Blaster"
- Grades 6 to 8
  - "Concert Tour"
  - "Shape Surveyor"

Additional Resources for Home Use:

To enhance your mathematics home learning environment, the following resources are suggested:

- "Helping Your Child Learn Mathematics" is a free booklet for parents to help strengthen children's mathematical skills and build positive attitudes toward mathematics. Activities are suitable for children in grades Pre-K to 5 and can be found at www.ed.gov/parents/academic/help/math/index.html.
- The website of the National Council of Teachers of Mathematics features *Figure This*, a resource consisting of middle school math challenges for families to explore. This feature also includes a "Family Corner" focusing on parental concerns and questions for all grade levels. This feature can be located at www.figurethis.org.

Classroom Happenings:

Your collaboration with your child concerning web-based resources continues to permeate their learning. Activities going on in their mathematics classroom include the following:

- group discussions concerning the interactive homework assignment
- written reflections concerning their collaboration with you

Your Comments:

To open our lines of communication even further, please share your feedback concerning any part of the parental involvement initiative. Feel free to suggest topics for future initiatives and to express concerns that can be addressed to help you nurture your home learning environment.

Please share your comments below and return to school by_____.

Sincerely,

SUGGESTIONS FOR FURTHER READING

Cesarone, B. (2000). Teacher preparation for the 21st century. *Childhood Education*, 76(5), 336–38.

Ertmer, R., Addison, P., Lane, M., Ross, E., & Woods, D. (1999). Examining teachers' beliefs about the role of technology in the elementary classroom. *Journal of Research on Computing in Education*, 32(1), 54–66.

Halpin, R. (1999). A model of constructivist learning in practice: Computer literacy integrated into elementary mathematics and science teacher education. *Journal of Research on Computing in Education*, 32(1), 128–35.

Haughland, S. W. (2000). What role should technology play in young children's learning? Part II: Early childhood classrooms in the 21st century: Using computers to maximize learning. *Young Children*, 55(1), 12–18.

International Society for Technology in Education. (2000). *National educational technology standards for students: Connecting curriculum and technology*. Eugene, OR: Author.

Kent, K. (2001). Are teachers using computers for instruction? *Journal of School Health*, 71(2), 83–84.

Lederman, N., & Neiss, L. (2000). Technology for technology's sake or for the improvement of teaching and learning? *School Science and Mathematics*, 100(7), 345–48.

Prawat, R. (1993). The value of ideas: Problems versus possibilities in learning. *Educational Researcher*, 24(7), 5–12.

Wenglinsky, H. (1998). *Does it compute? The relationship between educational technology and student achievement in mathematics*. Princeton, NJ: Educational Testing Service.

Wenglinsky, H. (2000). *How teaching matters: Bringing the classroom back into discussions of teacher quality*. Princeton, NJ: Educational Testing Service.

REFERENCES

Bitter, G., & Pierson, M. (2005). *Using technology in the classroom*. Boston: Pearson Education, Inc.

De La Cruz, Y. (1999). Reversing the trend: Latino families in real partnerships with schools. *Teaching Children Mathematics*, 5(5), 296–300.

Guastello, E. F. (2004). A village of learners. *Educational Leadership*, 61(8), 79–83.

Kerrigan, J. (2002). Powerful software to enhance the elementary school mathematics program. *Teaching Children Mathematics*, 8(6), 364–77.

National Council of Teachers of Mathematics (NCTM). (2000). *Principles and standards for school mathematics*. Reston, VA: Author.

Neiss, M. (2001). A model for integrating technology in pre-service science and mathematics content-specific teacher preparation. *School Science and Mathematics*, 101(2), 102–109.

Roblyer, M. D. (2003). *Integrating educational technology into teaching*. Upper Saddle River, NJ: Pearson Education, Inc.

Wright, R. T. (1999). Technology education: Essential for a balanced education. *NASSP Bulletin*, 83(60), 16–22.

Chapter Five

Playing Games and Solving Puzzles

Using games and puzzles in mathematical learning has also been proven to be effective in enhancing understanding and attitudes toward mathematics. MacDonald and Hannafin (2003) find games as a means to promote higher-order thinking skills as a result of increased verbal communication among participants.

In addition to being a motivational tool and confidence builder, the hands-on nature of playing games also provides opportunities for concrete learning (Shaftel, Pass, & Schnabel, 2005) that allows students to engage many times in more mathematics than when using traditional worksheets (Lee et al., 2004).

This chapter describes a parental involvement initiative designed to empower parents to help their children develop and enhance their mathematical thinking and attitudes through the use of games and puzzles. As in chapters 3 and 4, components of this chapter involve an invitation to parents along with explanations and related materials concerning the initiative's initial meeting, engagement workshop, follow-up session, and methods of maintaining the home-school mathematics connection.

ENCOURAGE YOUR PARENTS TO COME

As discussed in previous chapters, inviting parents into a community of learners nurtures productive collaboration. Following is the text for the letter of intent and inquiry that informs parents of the initiative's intent, requests their commitment to participate, and inquires about times that would best suit their busy schedules.

As in chapters 3 and 4, it is presented at this point for you to use when crafting your own letter of intent and inquiry or amend according to your specific situations. As described in previous chapters, personalize the invitation by having students design their own cover for it. Also included are a response form for you to use when gathering information about appropriate meeting times and a letter of announcement for you to use when beginning this initiative.

Letter of Intent and Inquiry

Dear Parents,

Children need to find mathematics experiences interesting and meaningful to reach their full mathematics potential. Using mathematical games and puzzles has proven to be successful in developing students' ability and attitudes toward mathematics. Your involvement with such a practice plays a vital role in supporting your child's mathematics education.

To empower you as partners with this aspect of your child's education, a parental involvement initiative titled "Playing Games and Solving Puzzles" has been designed for you and your child. Discussion on the purpose of games and puzzles in the learning of mathematics takes place along with the playing of specific games and solving of puzzles conducive to sound mathematical learning.

This initiative involves an initial meeting (1 hour), an engagement workshop (1 1/2 hours), home activities, a follow-up session (1 hour), and additional tasks to help maintain connections between the mathematics classroom and your home. The initiative *informs* you about the importance of parental involvement and the rationale behind specific teaching methods used in today's mathematics classrooms.

The initiative *engages* you and your child in playing games and solving puzzles involving the use of strategic thinking that extends into your home. Time for *reflection* on your learning experiences is facilitated as well between you and your child, among the parent community, and among the student community.

Please be aware that you are expected to attend the initial meeting and engagement workshop, engage in home activities with your child, attend the follow-up session, and maintain connections between the mathematics classroom and your home. Three meetings at school are required where one is attended only by parents and two are attended by both parents and their child.

If you wish to commit to this parental involvement initiative, please return the attached form. If you have more than one child, please arrange for each child to be represented by one family member.

We want to schedule this initiative in consideration of the busy schedules of all those involved. Therefore, please indicate on the attached form the time that would best suit you.

I look forward to working with you.

Sincerely,

The text of the response form that is mentioned in the above letter of intent and inquiry is presented below. As described in chapters 3 and 4, it fosters a sense of commitment, serves to investigate the most appropriate meeting times for your parent population, allows parents to indicate reasons they may not be able to participate, and serves as an organizational tool when keeping accurate records of attendance throughout the initiative.

Following the text of the response form is text for a letter of announcement. Spaces are provided for specific dates for the initial meeting, engagement workshop, and follow-up session. As mentioned previously, the text is shared with you for your direct use in creating your own response form and letter of announcement or as a framework to amend according to your personal preferences.

Response Form

Please return this form by _____.

We (commit, cannot commit) to the parental involvement initiative titled "Playing Games and Solving Puzzles."

If you cannot commit to this initiative, please explain why on the back of this form so we can try to partner with you in another way.

Name of Parent(s) and Other Family Members Committed to the Initiative (Each child must be accompanied by one family member.)

_____

_____

_____

Name of Child(ren) (with grade level) Committed to the Initiative (Indicate the family member above to partner with each child indicated below.)
_____
_____
_____

Parent(s) Signature(s): _____
_____
_____

Student(s) Signature(s): _____
_____
_____

Circle the most appropriate time for your family to commit to this initiative.
Weeknight beginning at 7:00 P.M. (Specify night) _____
Weeknight beginning at 8:00 P.M. (Specify night) _____
Saturday morning beginning at 10:00 A.M. _____
Saturday afternoon beginning at 1:00 P.M. _____
Sunday morning beginning at 10:00 A.M. _____
Sunday afternoon beginning at 1:00 P.M. _____

Letter of Announcement

Dear Parents/Family Members,

To empower you as partners in your child's learning of mathematics, the parental involvement initiative titled "Playing Games and Solving Puzzles" has been planned for you and your child. Your feedback concerning the most appropriate days and times for this event have been carefully considered, and the following dates and times reflect common responses.

Initial Meeting _____
(1 hour session for parents only)
Engagement Workshop_____
(1 1/2 hour session for parents and children)
Follow-up Session _____
(1 hour session for parents and children)

As noted in previous correspondence, the initiative *informs* you about the importance of parental involvement and the rationale behind specific teaching methods used in today's mathematics classrooms. The initiative *engages* you and your child in playing games and solving puzzles involving the use of strategic thinking that extends into your home. Time for *reflection* on your learning experiences is facilitated as well between you and your child, among the parent community, and among the student community.

Please be aware that you have committed to this initiative and are expected to attend the initial meeting and engagement workshop, engage in home activities with your child, attend the follow-up session, and maintain connections between the mathematics classroom and the home. All meetings start exactly on time, so please be punctual.

I look forward to working with you.

Sincerely,

INITIAL MEETING

As stated in chapters 3 and 4, the initial meeting should be attended by parents only due to its content. You should begin the meeting by presenting the initiative's goals and components using the information described in chapter 2 (Key Points to Share with Parents). As stated in previous chapters, this introduction allows parents to become knowledgeable of the initiative's structure and integrated features that serve to inform, engage, promote reflection, and maintain connections between the mathematics classroom and the home.

After this introduction, a discussion can follow with the use of the organized information discussed in chapter 1 (Key Points to Share with Parents) about the changes in mathematics teaching through the years, a constructivist foundation to teaching mathematics, the value of parental involvement, and the ways that parents can productively partner with their child at home to support mathematics education reform efforts. The discussion can then turn to the topic of using games and puzzles in mathematics learning.

As stated in chapter 4, if your parent populations have participated in a previous initiative described in this book, the introductory material from chapters 1 and 2 may be repetitious. If this is the case, just reminding parents of information covered in a prior initiative, and discussing the following information may be more appropriate. I leave that decision up to you and your individual settings.

Whatever direction you decide to take, the next step is to inform parents that game playing and puzzle solving positively affects mathematical learning at all grade levels, and it is important that they partner with teachers in using games and puzzles as a means to illuminating mathematical concepts, sparking strategic thinking, and establishing opportunities to develop logical reasoning skills (Martine, 2005).

Parents can play a key role in mathematical learning when they engage in games and puzzles with their children, share their own strategies, and encourage their children to do the same. Helpful information is organized below to facilitate your presentation and discussion about the rationale for using games and puzzles in mathematical learning.

Why Use Games and Puzzles?

- To highlight mathematical concepts
- To motivate
- To enhance attitudes toward mathematics
- To build confidence in themselves as a mathematical thinker
- To spark strategic thinking and logical reasoning skills

As with the information taken from chapters 1 and 2, this content should be used to create transparencies, PowerPoint presentations, and hand-outs as well so that parents have materials to refer to both during this initiative's events as well as at home.

Let the parents know they'll become engaged in playing games and solving puzzles that build conceptual understanding, computational skills, and problem-solving ability. Inform them that they will increase their awareness of game playing and puzzle solving as learning tools for mathematics.

As in previously described initiatives, allow a question-and-answer session to follow this discussion so that parents can contribute feedback and clarify any existing concerns. If questions from a small number of parents require in-depth responses and do not pertain to all those present, speak with these select parents after the workshop to avoid detaining others.

Before closing, remind parents that the engagement workshop involving both them and their children takes place two weeks later at the same time. As previously stated, I've found sending a reminder home to parents a day or two before this event to be helpful. A two-week interval between the initial meeting and engagement workshop gives parents enough time to review the material presented at the initial meeting and remain vibrant about the initiative's intentions.

ENGAGEMENT WORKSHOP

As in the initiatives described in previous chapters, you are at a point now where you have given your parents the background information they need to understand the rationale behind current mathematics teaching methods, specifically the use of games and puzzles, and the value of their involvement in their child's mathematics education.

The engagement workshop can in turn take on a different environment that is quite active and filled with mathematical reasoning while participants involve themselves in three activities, namely, "Show Your Age," "Attribute Line Up," and "Pattern Block Reasoning Puzzles." To create an appropriate environment, this workshop needs to take place in a room with sufficient tables for game playing, puzzle solving, and discussion.

You may find that using all three of the activities causes you to run over the allotted time for this workshop, especially if it is your first time implementing this type of experience. I therefore suggest choosing two out of the three activities. Have the third one on hand if time allows. You can then judge better for future initiatives whether two or three activities works best for you.

Show Your Age

Show Your Age is a simple game that can be played by children and parents of all grade levels. It is an excellent way to spark conversation and illuminate multiple correct answers that in turn builds confidence and fluency with multiple representations of number.

Begin by asking the entire group "How old are you?" Elicit a response from a participant, and record it on a surface that they can all see. An example of a response would be "I am nine years old." Next pose the question "How many ways can we express nine?" Record the participants' responses such as 9, nine circles, 8 + 1, 11 − 2, 3 x 2 + 3, 9 percent, nine cents (one nickel and four pennies), and so on.

Once the group is motivated and clear about what is being asked of them, pass out paper and allow each child and their parent to brainstorm together for approximately ten minutes on finding as many ways as they can to represent both the child's and the parent's age. Offer a prize (a no homework pass, a dress down pass, a giant candy bar, etc.) to the winning parent-child team.

Have someone on hand who can gather the papers and decide on a winner based on not necessarily the quantity of answers but rather the variety of answers. I say this because I have had instances where team responses ranged from involving just addition to responses that incorporated different operations, pictures, and money just to name a few.

Whatever the grade level(s) you are working with, encourage everyone from the beginning to think of as many different ways to represent their ages. Doing this allows you to point out while sharing responses the many connections made to the curriculum.

As this brainstorming session comes to a close, bring the entire group back together for some dialogue (approximately five minutes) concerning the following questions:

- Who took the lead?
- Did someone represent a number in a way that you didn't think about?
- How?
- How did you react to different approaches to this task?
- Would you change anything the next time you do this task?

At the close of this discussion, gather the papers for review and announce the winner at the end of your session together. A giant candy bar or a no homework pass for the winner has proven for me to be an appropriate prize.

Attribute Line Up

Attribute Line Up involves a die and a twenty-four-piece attribute set involving three shapes (square, circle, and triangle) that are two sizes (large and small) and four colors (red, blue, green, and yellow), and a set of thirty difference cards (one set for grades Pre-K to 2, and another set for grades 3 to 8). A template for the attribute pieces can be found at the website known as Teachervision. The website address is indicated below.

- www.teachervision.fen.com/geometry/printable/44607.html?detoured=1

This template can be used with your students prior to the engagement workshop so that the children and their parents each have an attribute set for this part of the workshop as well as for home use. Copy the template onto durable cardstock and have your students cut the template page into squares so that each square contains a shape.

For the younger students, you may find it better to cut the squares out yourself or send the page home with a letter to the parents explaining how to cut out the squares. If this is the case, allow enough time for the parents to cut out the squares and return them to you prior to the engagement workshop.

The students then need to color the shapes so that they have a large and small version of each shape in each of the four colors. For example, they should have a large and small blue triangle, a large and small red triangle, a large and small green triangle, and a large and small yellow triangle. The same holds true for the remaining shapes. Once all of the shapes have been cut and colored, store them in separate Ziploc plastic bags so that they can be distributed easily at the engagement workshop.

As stated in chapter 3, there are instances when a website that has functioned in the past becomes nonexistent for various reasons and can cause the creation of paper attribute pieces to be a challenge. If you find yourself in a situation where the website I mentioned earlier is not functioning, I recommend two alternatives.

The first is to purchase an inexpensive plastic attribute piece template from ETA/Cuisenaire Company (www.etacuisenaire.com). Using this template, you can draw the attribute pieces described earlier onto a piece of paper and then copy them onto cardstock.

The other alternative is to draw the attribute pieces yourself. I have found the drawing toolbars for word processing programs such as Microsoft Word to be very useful for such a task. You can then copy the attribute pieces onto cardstock. If you choose either of these two alternatives, the actual attribute pieces are cut out and colored as opposed to the squares containing the shapes as is the case when using the online template.

Each team also needs to have a set of thirty difference cards, so having the students make these card sets before the engagement workshop as well saves time. For grades Pre-K and K, I have found that sending a note home to parents requesting that they make the card set to be very helpful. Or you can make the card sets yourself.

The thirty difference cards for grades Pre-K to 2 are a set of index cards consisting of ten cards that each say "color," ten other cards that each say "shape," and ten cards that each say "size." For grades 3 to 8, the thirty-card set consists of ten cards each saying "1 difference," ten cards each saying "2 differences," and ten cards each saying "3 differences."

To begin an explanation and demonstration of how this game works, distribute to each child and their parent a die, a bag of twenty-four attribute pieces, and a set of thirty difference cards. Discuss as a whole group the attributes of the pieces (shape, color, and size). Asking the group to describe how the pieces differ from one another can help elicit that the set contains twenty-four pieces consisting of three shapes, two sizes, and four colors. Now each team (child and parent) is ready at various degrees of difficulty to play Attribute Line Up.

Each team mixes the attribute pieces up and divides them in half so that each player has twelve pieces. The set of difference cards is shuffled and placed in the middle of the two players who should sit opposite each other. Each player rolls a die, and the player with the highest roll goes first by choosing one of their attribute pieces and placing it in front of them. The other player then chooses one of their attribute pieces and places it in front of them. The first player then takes another turn by choosing a card from the deck of difference cards.

For grades Pre-K to 2, suppose the card picked says "color." That would indicate that the player needs to choose one of their attribute pieces that is a different color from the first piece they chose. The player then places that piece next to the first piece forming a line of two attribute pieces. The card they chose should be placed face down in a separate pile.

Play then turns to the next player who chooses a card from the deck as well. Suppose this player picks a card that says "shape." In this case, that player needs to choose an attribute piece of theirs that is a different shape from the first piece they chose. The player then places that piece next to their first piece also forming a line of two attribute pieces. The card they chose is placed face down in the separate pile that the other player formed when it was their turn.

Play continues in the same fashion with each player picking a card and adding to their line of attribute pieces with each attribute piece being different from the one before it in only one way (color, shape, or size). Suppose, on the third round of play, a player chooses a card that says "size." That player would then need to choose an attribute piece of theirs that is different from the last piece in their line by size.

Players must understand that when adding to their line of attribute pieces, the next piece placed is always different from the piece before it by color, size, or shape. The piece may also end up being different in other ways as well from the piece before it, but for this grade level, focusing on the difference indicated on the card is sufficient.

For example, if a player has a small red triangle as the last piece in their line and they pick a card that says shape, choosing a big red square is perfectly alright. This piece happens to be different from the small red triangle by both size and shape. But as long as they change the shape, the choice is considered correct. In the older grades, more specificity is appropriate allowing the only correct choices to be the small, red square and the small, red circle. You may even choose to require such specificity in grades 2 and 3 as well.

For grades 3 to 8, the game is started in the same way. The only difference is the card set. If a player chooses a card that says "1 difference," that player needs to choose an attribute piece that is different from the one before it in only one way (size, shape, or color). For example, if the last piece in a player's line is a big, yellow square, then one correct choice would be a big, yellow circle with the one difference being shape.

If a player chooses a card that says "2 differences," that player needs to choose an attribute piece that is different from the one before it in two ways (size and shape, size and color, or shape and color). For example, if the last piece in a player's line is a big, yellow square, then one correct choice would be a big, red circle with the two differences being color and shape.

If a player chooses a card that says "3 differences," that player needs to choose an attribute piece that is different from the one before it in three ways (size, shape, and color). For example, if the last piece in a player's line is a big, yellow square, then one correct choice would be a small, red triangle.

As the game progresses, there will come times when a player is unable to line up an attribute piece of theirs with the others. When that happens, play turns to the other player. The first player to line up every one of their attribute pieces wins the game.

Once you notice that the teams have finished playing (usually takes about fifteen minutes), bring the group together to discuss what happened (approximately five minutes). Pose the following questions to facilitate reflection as a whole group.

- Was this game enjoyable?
- Was this game challenging at points? If so, how?
- Would you strategize differently the next time? If so, how?

During this reflection session, I have found parents and children to be very positive about this game. They've expressed "feeling their heads reason" throughout the game. Some feedback I wish to share with you comes from parents and children in grades Pre-K to 2. They voiced that, when playing again, they would try to choose pieces in a way that would leave them with pieces of different sizes, shapes, and colors on reserve for as long as they could.

For example, if they needed to pick a piece different from a large, yellow triangle by size, they might choose a small, green circle as opposed to a small, yellow triangle so that they could keep different colors and shapes on reserve since they already used a yellow and a triangle on a previous turn. Such brainstorming is an example of parents and children productively collaborating as they strategize together.

Pattern Block Reasoning Puzzles

The last activity to take place at this engagement workshop involves Pattern Block Reasoning Puzzles and uses pattern block pieces, puzzle grids, and pencils. These puzzles engage parents and children in reasoning about the position of pattern blocks in a grid based on a given number value.

I have found these puzzles useful in acquainting parents with yet another manipulative used by their children in the mathematics classroom and to cultivate a learning environment where participants are connecting concrete learning of fractional concepts with computational skill building and higher order thinking.

To begin solving these puzzles, parents and children need to explore together the relationships existing among pattern blocks, specifically, the hexagon, trapezoid, parallelogram, and triangle. Therefore paper (cardstock) pattern blocks need to be prepared beforehand so that parents and children can use them both during the engagement workshop and at home. Printable pattern blocks can be found on the website below.

- http://mason.gmu.edu/%Emankus/Handson/manipulatives.htm

You need not print all of the pages on this website. Select the pattern block pages. Print the hexagon page in yellow, the trapezoid page in red, the parallelogram page in blue, and the page containing triangles in green. Cut out the pattern blocks, and place them in Ziploc baggies so that each bag contains at least one yellow hexagon, two red trapezoids, three blue parallelograms, and six green triangles. If time does not allow you to do this yourself, send the papers home, and through a letter ask some of your parents to cut them out. You can later sort them into the bags or have your students do this task if they are able.

Again, stated previously, there are instances when a website that has functioned in the past becomes nonexistent for various reasons and can cause the creation of paper pattern blocks to be a challenge. If you find yourself in a situation where the website I mentioned earlier is not functioning, I recommend two alternatives.

The first is to purchase an inexpensive plastic pattern block template from ETA/Cuisenaire Company (www.etacuisenaire.com). Using this template, you can draw the pattern blocks described earlier onto a piece of paper and then copy them onto the appropriate color cardstock.

The other alternative is to draw the pattern blocks yourself. I have found the drawing toolbars for word processing programs such as Microsoft Word to be very useful for such a task. Choose the hexagon from the standard shapes (AutoShapes) available and double its size. Copy this hexagon onto the page until you have a full page of hexagons. Then print out four copies of this page.

Take one of the pages and divide each of the hexagons in half to form a page of trapezoids. Take another page and divide each of the hexagons into thirds to form a page of parallelograms. Take another page and divide each of the hexagons into sixths to form a page of triangles. If you are like me, you may find it easier to make the triangle page first using a pencil and then go back and erase some lines to form the parallelograms.

You can then copy the hexagon, trapezoid, parallelogram, and triangle pages onto the appropriate color cardstock. If you choose either of these two alternatives, the pattern blocks would then be separated into bags as described earlier.

Begin by introducing the shapes using an overhead projector and transparency pattern blocks. Distribute the bags of pattern blocks and ask parents and children to explore the shapes (approximately five minutes). Ask them to cover the yellow hexagon with red trapezoids, then with blue parallelograms, and finally with green triangles. As this exploration time comes to a close, pose the following questions:

- How many red trapezoids covered the yellow hexagon?
- How many blue parallelograms covered the yellow hexagon?
- How many green triangles covered the yellow hexagon?
- How many green triangles cover a red trapezoid?
- How many green triangles cover a blue parallelogram?

These questions set the stage for discussion about the fractional relationships among the pieces that is needed when solving the pattern block puzzles. For example, the green triangle can be seen as one-sixth of the yellow hexagon, one-half of the blue parallelogram, and one-third of the red trapezoid. The blue parallelogram can be seen as one-third of the yellow hexagon, and the red trapezoid can be seen as one-half of the yellow hexagon. When describing the yellow hexagon, one can also say it is six times the size of one green triangle, three times the size of one blue parallelogram, or two times the size of one red trapezoid.

The following pattern block puzzle (see figure 5.1) is appropriate for grades Pre-K to 2 and should be used to demonstrate to all grade level groups the manner to solve the puzzles. It is a puzzle that involves the pattern block triangle, parallelogram, and trapezoid.

Explain to parents and children that to solve this puzzle, one first needs to determine the value of the pattern block pieces using the relationships among the pieces they discovered previously along with the given num-

Puzzle

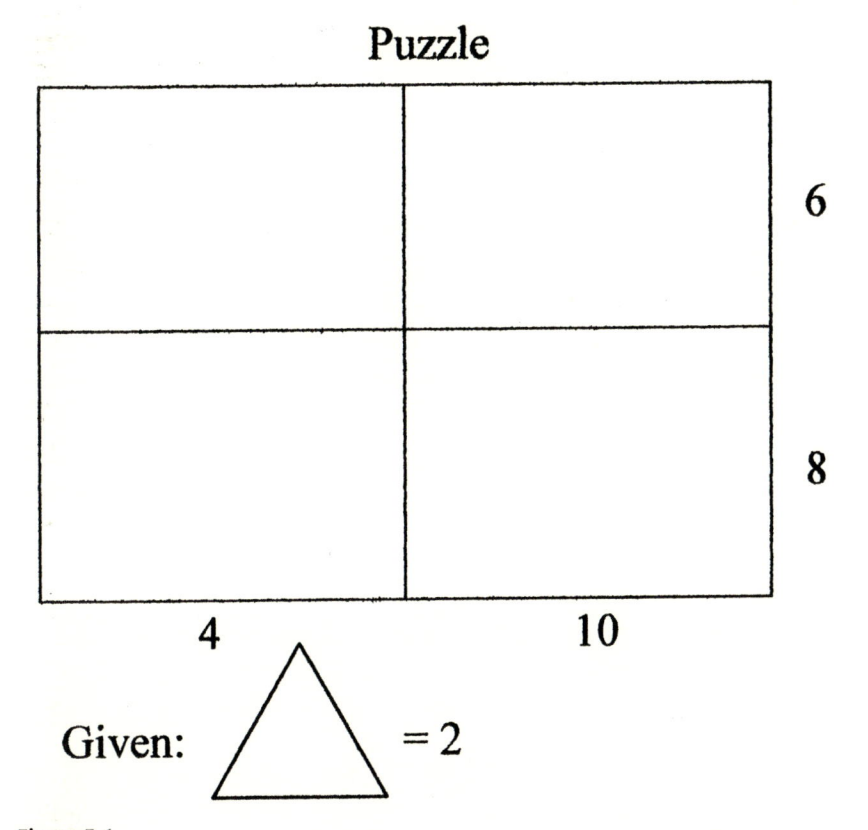

		6
		8

　　4　　　　　10

Given: △ △ = 2

Figure 5.1.

ber value for the green triangle. For example, in this case the green triangle is given a value of two. Elicit from the group that the blue parallelogram would then equal four because two green triangles cover a blue parallelogram (2 + 2 = 4). Elicit from the group that the red trapezoid equals six because three green triangles cover a red trapezoid (2 + 2 + 2 = 6).

Once the number value of each pattern block piece is established, inform the group that they then need to place the pattern block pieces into the grid so that when the number values of the pieces are added horizontally and vertically, the sums are equal to those numbers around the border of the grid. Let them know that they can use a pattern block more than once, and allow them time to strategize (approximately five minutes).

I suggest telling the parents and children to write the number values onto the shapes in pencil before placing the pieces onto the grid so they can focus on the number values. When the puzzle is solved, the pencil markings can be erased and the pattern block pieces reused for another puzzle.

As a whole group, discuss the solution (see figure 5.2) and the methods used to arrive at them.

Highlight the need to establish a horizontal row or a vertical column where the pattern block pieces are the same. In the puzzle above, the first column contains only green triangles (2 + 2 = 4) and leads to the appropriate placement of the other pattern block pieces. The top row contains the triangle and the parallelogram (2 + 4 = 6) while the second column contains the parallelogram and the trapezoid (4 + 6 = 10).

Following are other puzzles and their solutions (see figures 5.3 and 5.4) that can at this point be solved by parents and children (allow approximately ten minutes). I suggest using the puzzle in figure 5.3 for grades 3 to 5 and the puzzle in figure 5.4 for grades 6 to 8. I have included the solutions to each puzzle as well for your use. Please note that the puzzle for grades 6 to 8 also involves the hexagon. For this puzzle, you should review that six triangles can cover a hexagon. Since the

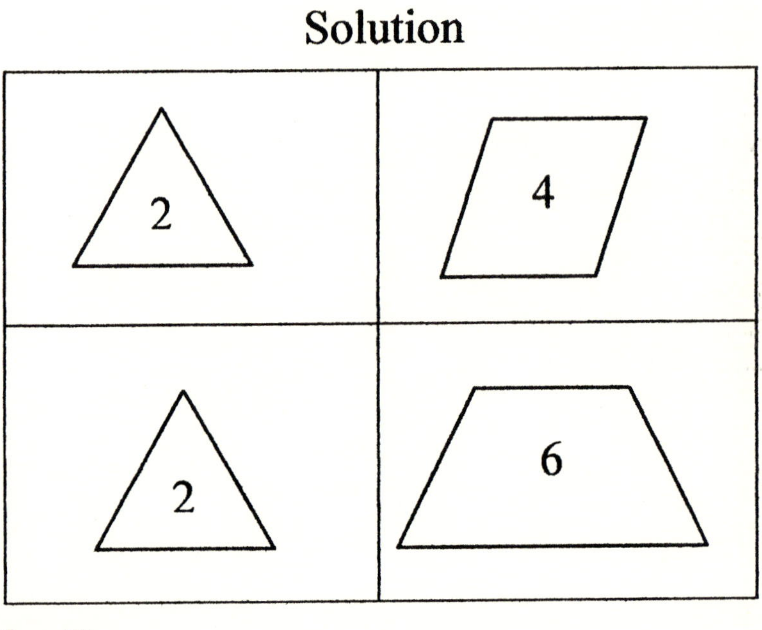

Solution

Figure 5.2.

Puzzle

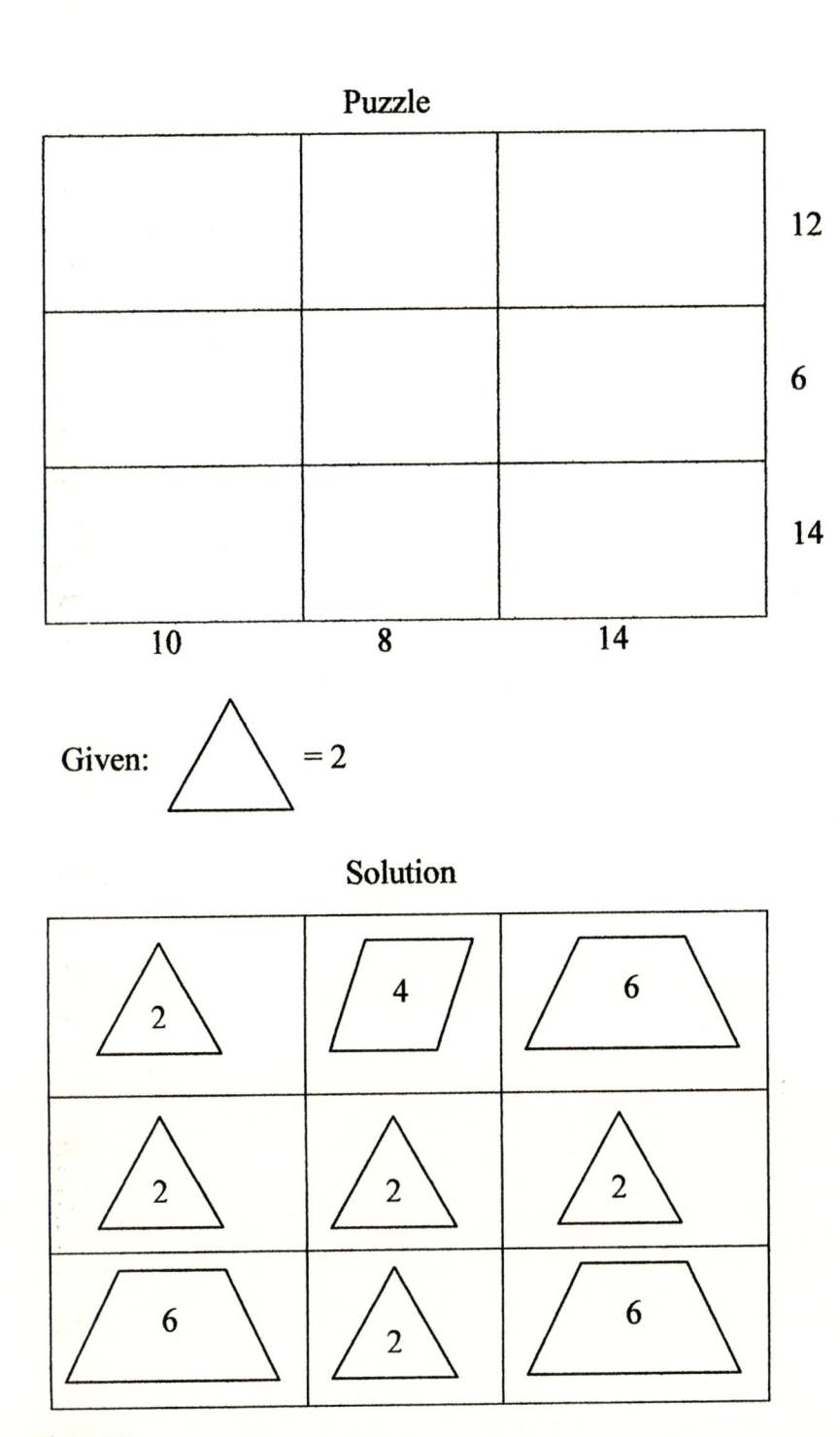

			12
			6
			14
10	8	14	

Given: △ = 2

Solution

△ 2	◇ 4	⬠ 6
△ 2	△ 2	△ 2
⬠ 6	△ 2	⬠ 6

Figure 5.3.

Puzzle

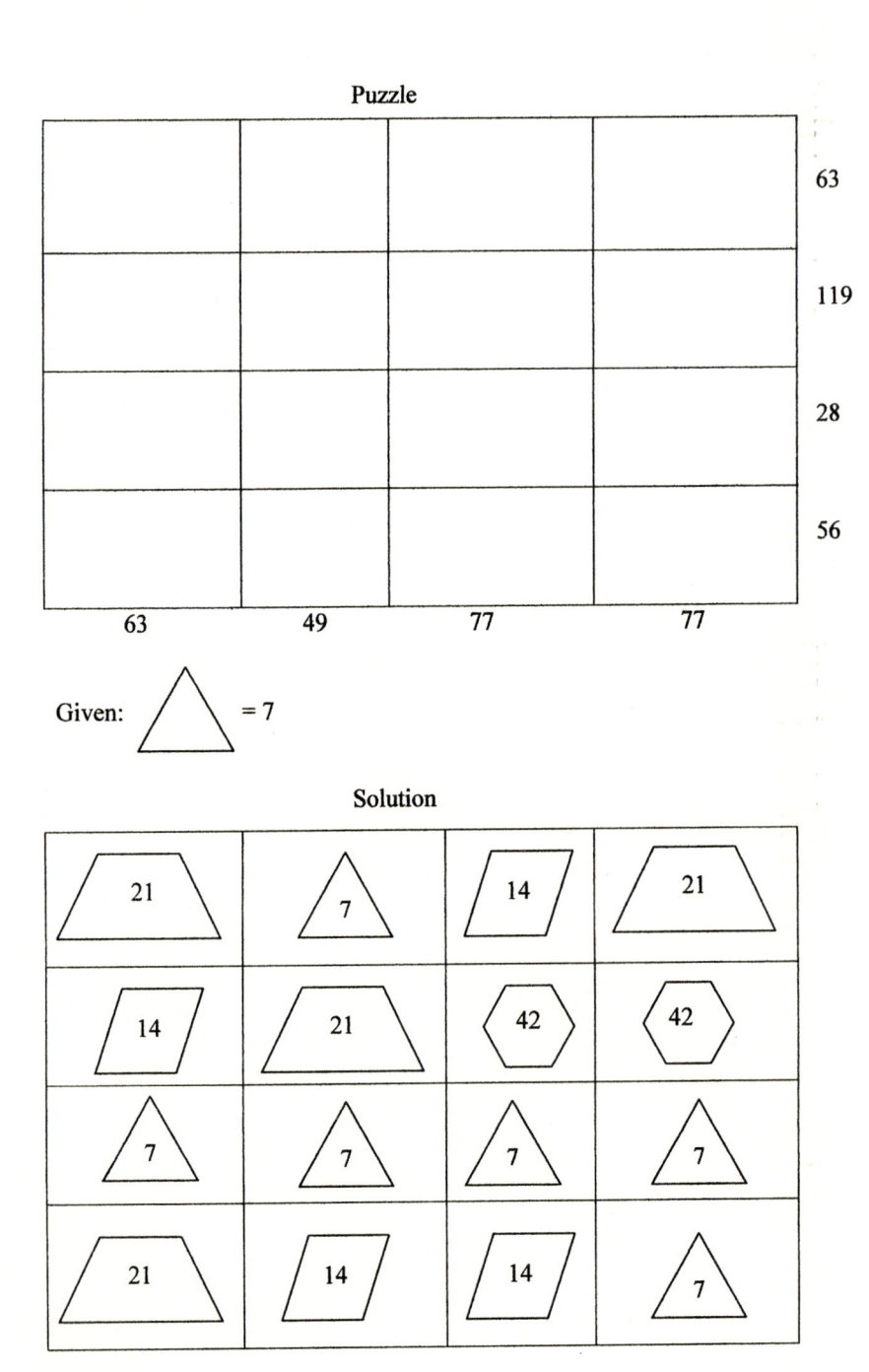

				63
				119
				28
				56
63	49	77	77	

Given: △ = 7

Solution

21	7	14	21
14	21	42	42
7	7	7	7
21	14	14	7

Figure 5.4.

triangle has a value of seven, the hexagon has a value of forty-two (7 + 7 + 7 + 7 + 7 + 7 = 42).

You can make your own puzzles as well. Creating them yourself gives you the opportunity to involve the basic facts and computational skills in your puzzles that are relevant to the grade level(s) you are working with. Parents are then resources to you who can strategize with their children in a task that merges conceptual understanding of the fractional relationships existing among the pattern block pieces with computational skills.

To create your own puzzles, start out with a blank grid, assign a value to one of the pattern block pieces, and determine the other values. I suggest using two-by-two grids for grades Pre-K to 2 and increasing the size according to the ability levels of your children.

Arrange the pieces in the grid, remembering that there needs to be a row or column that contains the same shape. From that arrangement stems the sums for your puzzle that are placed around the border of your grid.

It is also a nice idea to give a number value to a pattern block piece other than the green triangle as I have done in the home activity for grades 6 to 8 that I share later in this section. In that example, a value of forty-eight is given to the yellow hexagon that causes one to think about various operations (addition, subtraction, multiplication, and division) to arrive at a value of eight for the green triangle, a value of sixteen for the blue parallelogram, and a value of twenty-four for the red trapezoid. Using decimal and fractional values for a pattern block piece can increase the challenge as well.

Once you notice that the teams have finished solving their puzzle, bring the group together to review their solutions (approximately ten minutes). Pose the following questions to facilitate reflection.

- Was this puzzle challenging? If so, how?
- What strategies did you use?
- Were there different ways you could have arrived at your solutions?

During this reflection session, I have found parents and children to be very motivated. They have expressed an appreciation for the challenge and for the opportunity to work collaboratively with their child.

Upon review of parents' evaluation sheets for this initiative, it was noted that the parents saw how mathematics concepts, computational skills, and higher order thinking are integrated in mathematical tasks. They also realized how they can take on more of a role in their child's mathematical learning that goes beyond just drilling basic facts and checking if homework is complete.

You may have noticed that this initiative differs from the previous initiatives' format in that time for small group reflection does not take place at this time. In this initiative, small group discussions take place only in the follow-up session. I have found the amount of activity and discussion to be appropriate since parents and children have been actively involved in working as a team and discussing their experiences as a whole group at three different points during the engagement workshop.

Take the remaining time (approximately fifteen minutes) to both reinforce and extend this workshop experience by setting the stage for learning at home. As in chapter 4, a cover letter that is appropriate for everyone and grade level home activities that reinforce and extend the engagement workshop experience are provided for you at this point. You may use the content as is or amend it according to your specific situations as you develop your materials for a home learning experience.

In closing the engagement workshop, distribute and review these materials (including sets of paper attribute pieces and pattern blocks) with everyone so that the parents and children are clear about what needs to be done as they engage in doing mathematics together at home using games and puzzles.

As stated in previous chapters, instruct parents to collaborate with their children in the same manner they did during the engagement workshop. Remind them to not do all of the telling, but rather to communicate and explore together. Most important, emphasize the importance of just having fun at home with mathematics.

After the activities have been completed, parents are to respond to a reflection paper concerning the experience. This reflection paper is included for your use on the following pages. Inform the parents that this reflection paper and their responses must be brought to the follow-up session so that discussion as a community of mathematics learners using games and puzzles can be facilitated.

Cover Letter

Dear Parents and Students,

Presented to you in this letter are activities for you to do at home that both reinforce and extend our learning from the engagement workshop. The activities relate to the games and puzzles we explored together during the engagement workshop. Remember to again share your thinking as you play and strategize together. Parents, remember that if you do all of the telling this is not a productive experience.

After you complete each of the activities, complete the attached reflection paper as well. Our follow-up session takes place on _____ at _____ so we can discuss your experiences. Please bring the reflection papers with you. This session (approximately one hour) starts exactly on time, so please be punctual.

I look forward to discussing with you your reactions to the activities. As you work together, keep in mind the importance of exploring mathematics together, sharing your thinking, and just having fun.

Sincerely,

Home Activities

- Show Your Age
  - Use the age of a family member other than the one you worked with at the engagement workshop and represent (show) that age on a piece of paper in as many creative ways as you can.
  - At the top of the paper, write the age of your family member as follows:
  - My family member is my _____ and is _____ years old.
  - Underneath that sentence, write/draw all of your representations (ways of showing that number). Bring this paper with you to our follow-up session.
- Attribute Line Up
  - Play a game of Attribute Line Up as we did at the engagement workshop. Use the attribute pieces and the game cards given to you at that

120   Chapter Five

workshop. After someone wins, look at the winner's line up of attribute pieces. On a piece of paper, draw a picture of that winning line up of attribute pieces. Talk with your partner about how each piece is different from the other. Then, underneath your picture, write about how each piece is different from the other, or have your partner write down what you say (for grades pre-K to 2). Bring this paper with you to our follow-up session.
- Pattern Block Reasoning Puzzles
  - Solve the puzzle on the following page for your grade level. Remember that you need to first get a column or row where the shapes are the same. On a separate piece of paper, write about what you were thinking as you got each answer, or have your partner write down what you say (for grades Pre-K to 2). Bring the puzzle and paper on which you explained your thinking to our follow-up session.
    - Grades Pre-K to 2

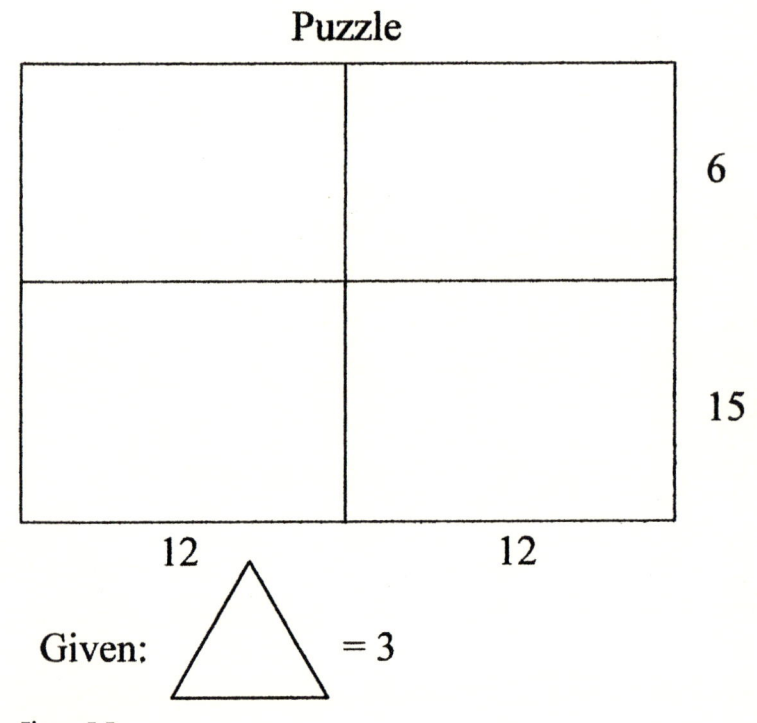

Puzzle

		6
		15
12	12	

Given: △ = 3

Figure 5.5.

Playing Games and Solving Puzzles

- Grades 3 to 5

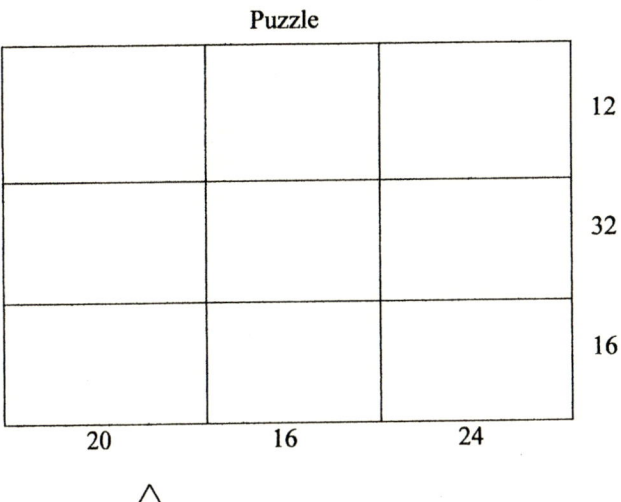

Puzzle

			12
			32
			16
20	16	24	

Given: △ = 4

Figure 5.6.

- Grades 6 to 8

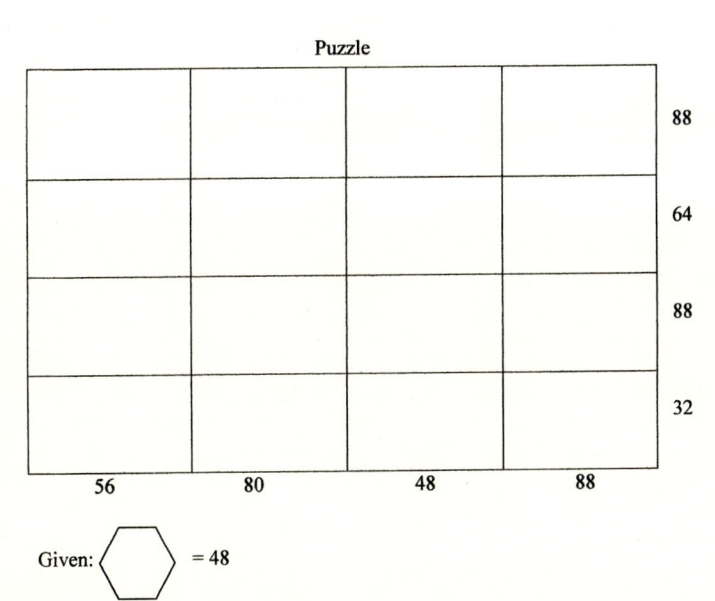

Puzzle

				88
				64
				88
				32
56	80	48	88	

Given: ⬡ = 48

Figure 5.7.

Reflection Paper

Answer the following questions as you work together on each of the tasks. Bring your written responses with you to our follow-up session.

- In what ways was Show Your Age most helpful?
- In what ways was Attribute Line Up most helpful?
- In what ways was the Pattern Block Reasoning Puzzle most helpful?
- What did you find most interesting about using the games and puzzles?
- How has this experience impacted your family?

FOLLOW-UP SESSION

As stated in previously described initiatives, allow two weeks before holding the follow-up session so that parents have enough time to complete the activities with their child in a manner that is not rushed. I've found that sending a reminder letter home a week before helps alert parents to this upcoming event. The main goal of this session is to share families' experiences with the games and puzzles and reflect as a community of mathematical learners.

Small group discussion was not a feature of the engagement workshop as it was in previously described initiatives due to the structure of this initiative. Explanations, demonstrations, and actual game playing and puzzle solving did not allow for time to be spent in small group discussions. Rather, large group discussions were a more efficient use of time at the engagement workshop, leaving this follow-up session as a perfect time for small group discussions.

Place the parents and children into groups of two to three families to share their work concerning the three tasks using their responses to the questions of the reflection paper (approximately twenty minutes) they took home with them at the close of the engagement workshop. Designate two people as was done in previous initiatives to record responses and report on them. When you notice that all has been accomplished within the groups, bring the entire group together for a whole group discussion and record their responses.

As stated previously, past experience has informed me that such reflection empowers parents to support each other as a community of mathematics learners. In addition to the parent comments I shared previously from other initiative's follow-up sessions, other comments during this initiative include the following: "My child loves games, and now I can connect game playing with math learning" and "This is a win-win situation: my child is happy because it's fun; I'm happy because he's doing math."

In bringing this session to a close, take the remaining time to both discuss the ways that the home-school connection is maintained and emphasize how game playing and puzzle solving connects to current mathematics curricula by building conceptual understanding, computational skills, and problem-solving ability.

Explanations of the interactive homework assignment, classroom follow-up techniques, and the interactive newsletter that serve to maintain the home-school connection can be found at the end of the follow-up section of chapter 3. As stated there, the content can be used for transparencies, PowerPoint presentations, and hand-outs.

As you discuss the value of this initiative's games and puzzles and their connection to the *Principles and Standards for School Mathematics* (NCTM, 2000), highlight how the game Show Your Age emphasizes representation of numbers in ways that allows for deeper conceptual understanding of numbers, practice with computational skills, and reasoning about mathematical situations. An example of the last curriculum connection is when problem-solving situations are created where the answer to the situation is the number (age) being represented.

Point out that the game Attribute Line Up requires children to sort, classify, and use reasoning skills to appropriately place their pieces. When reviewing the Pattern Block Reasoning Puzzles, stress how the puzzles place participants in situations where they use fractional relationships among the pieces and whole number operations to strategize and eventually arrive at solutions.

MAINTAINING THE CONNECTION

To ensure continuity of the strides you have made during this initiative, your students and their parents need to remain connected. In this section, I share, as I did in previous chapters, ways to accomplish this through an interactive homework assignment, classroom follow-up techniques, and an interactive newsletter designed to keep the home-school connection alive.

Interactive Homework

As stated previously, interactive homework engages both the child and his or her parent in a mathematics task where input from both parties creates the final product. It gives parents the opportunity to communicate not only with their child about his or her reasoning but also with their child's teacher about the progress made with the assignment and any existing concerns.

The interactive homework assignment on the following pages engages parents and children in building upon fractional relationships established with the Pattern Block Reasoning Puzzles. You'll notice that the tasks are similar to those involved in the initiative discussed in chapter 3 that used tangrams. As was done in that initiative, designs are created and costs for them computed. Instead of designing with tangrams though, pattern blocks are used.

The format consists of (1) a "Dear Math Learning Partner" section where the parent is invited to work with their child on a mathematics task; (2) a "Take a Look" section to give necessary background information; (3) a "Let's Try This Together" section to facilitate a cooperative effort between the child and their parent; (4) a "Can We Do This Another Way?" section to promote deeper experiences with creating a Pattern Block Reasoning Puzzle; and (5) a "What Do You Think?" section to provide parents with the opportunity to give feedback about the assignment and allow teachers to assess this component of the initiative.

As stated previously, the content is presented for you to use directly or amend according to your personal preferences and situations as you craft an interactive homework assignment for your student and parent populations. The content presented below for each of the sections of the interactive homework assignment can flow after each other.

The only section that should appear alone on a separate page is the parent feedback section so that it can be detached and reviewed separately from the actual assignment.

Please note that included in this assignment needs to be a set of paper pattern blocks for the parents and children to use at home. These can be the same sets they used during the engagement workshop and at home, sets that the students create in school, or sets that the parents and children can cut out at home from printouts that you send home on appropriately colored paper (green for triangles, blue for parallelograms, red for trapezoids, and yellow for hexagons).

Dear Math Learning Partner,
Here is some homework we can do together about puzzles. I hope you like it. This assignment is due _____.
   Thanks for working with me.

Your Math Learning Partner,

_____
Student's Signature

Take a Look

Remember when we worked with Pattern Block Reasoning Puzzles? We used green triangles, blue parallelograms, red trapezoids, and yellow hexagons. Let's use our pattern blocks to review what we did at the engagement workshop.

   We found out that a yellow hexagon can be covered by six green triangles, three blue parallelograms, or two red trapezoids. We also found out that two green triangles cover a red trapezoid and two green triangles cover a blue parallelogram.

   We talked about how the green triangle is one-sixth of the yellow hexagon, so a yellow hexagon is six times the size of one green triangle. The blue parallelogram is one-third of the yellow hexagon, so a yellow hexagon is three times the size of one blue parallelogram. The red trapezoid is one-half of the yellow hexagon, so a yellow hexagon is two times the size of one red trapezoid.

   We needed to find out about these relationships among the pattern block pieces to solve the pattern block puzzles. Now we are going to use what we know to create designs and compute the cost of our designs.

Let's Try This Together

Let's each make a design with our pattern block pieces on a separate piece of paper. We can use as many pieces as we wish. Once we are pleased with our designs, we need to glue the pieces onto the paper. Now we are going to use a certain money amount for the green triangle to figure out the total cost for each of our designs.

- If we are in grades Pre-K to 2, the green triangle costs two cents.
- If we are in grades 3 to 5, the green triangle costs fifty-three cents.
- If we are in grades 6 to 8, the green triangle costs $1.89.

Now let's find out how much our designs cost. We need to show our work on the paper we have our design on. We need to explain our thinking on a separate piece of paper.

To play a little game, we now need to take four small pieces of paper and write a money amount on each of them. One should say $0.75, another $2.50, another $35.73, and another $50.00. Let's put these pieces of paper in a hat and pick out one. Whoever has the design that costs closest to the amount on the piece of paper is the winner.

Can We Do This Another Way?

Can we find the cost of our design another way? If so, what can we do differently? If another way isn't possible, explain why there is only one way to find the cost.
Explain.

_____
_____
_____
_____

Can we each change the cost of the green triangle in some way so that the cost of our design comes closer to the cost we picked out of the hat? If so, how can we change it? If not, explain why we can't change it.
Explain

_____
_____
_____
_____

Parent Feedback: Please let my teacher know how we did together with this assignment.

_____ My child did well. The idea of this assignment was understood, and its tasks were successfully completed.

_____ My child needed help. The idea of this assignment seems to be understood, and its tasks seem to be successfully completed.

_____ My child needs instruction. The idea of this assignment was not understood, and its tasks were not successfully completed.

Comments:

_____

_____

_____

Parent's Signature:

_____

Classroom Follow-up Techniques

As stated in previous chapters, reflecting on the experience of working with their parents allows students to review the components of the initiative, organize their thoughts, and give insight into what has been successful and what needs to be improved. In this section, I discuss ways that you can strengthen the parental involvement sparked by this initiative's efforts through use of group discussion and written reflection in the mathematics classroom about the use of games and puzzles.

As stated in previous chapters, to maximize the potential of the interactive homework assignment and foster communication among students about learning at home with their parents, pair the students so that they can share their responses on the interactive homework assignments and offer feedback to each other. Students should not be paired until you have reviewed their responses, detached the parent feedback sheet, and given assistance where needed so that all have valuable feedback to share.

To initiate the discussions, engage the pairs in responding to each other about the following questions:

- Show each other the designs you and your parent created.
- Show each other your answers and explanations.
- What was most enjoyable about doing this assignment with your parent?
- What was most challenging about doing this assignment with your parent?

Once you notice that your students have adequately discussed the questions, have them share their responses in a whole group setting. This maximizes their interaction as a community of learners who are growing with their parents as partners in mathematical learning.

In addition to group discussions, as stated previously, I have found it helpful to engage the students in writing letters to their parents. These letters can in turn be sent home to help nurture home learning environments and guide your efforts in planning future initiatives. The format for these letters is included in chapter 3 in the section concerning classroom follow-up techniques.

For the younger students whose writing skills may still be developing, have these students draw a picture that depicts their feelings about doing mathematics with their parent.

Interactive Newsletter

Even parents who may not have participated in this initiative may be intrigued by reading the newsletter and in turn choose to participate in future initiatives.

The content presented below is for your use while crafting your own interactive newsletters. I suggest placing your content into actual newsletter formats available through software programs such as Microsoft Word, Publisher, and PrintMaster.

Math Notes

Dear Parents,

This interactive newsletter serves to review the events of our parental involvement initiative titled "Playing Games and Solving Puzzles," provide additional support for learning at home with games and puzzles, and share classroom happenings related to this topic.

Newsletter contents:

- Workshop Events Revisited
  - Initial Meeting
  - Engagement Workshop
  - Follow-up Session

- Game and Puzzle Information
- Additional Resources for Home Use
- Classroom Happenings
- Your Comments

Workshop Events Revisited:
Our parental involvement initiative has involved three events (initial meeting, engagement workshop, and follow-up session). The initial meeting informed us about the following:

- changes in mathematics teaching through the years;
- a constructivist foundation to teaching mathematics;
- the value of parental involvement;
- ways that parents can productively collaborate with their children at home to support mathematics education reform efforts;
- the value of game playing and puzzle solving in mathematical learning.

Our engagement workshop actively involved us in playing games and solving puzzles with the children. We strategized and explored each other's thinking while engaging in tasks that involved conceptual understandings of mathematical ideas and computational skills. We shared our findings and experiences and left the workshop with games and puzzles to do at home.

Our follow-up session was filled with reports from parents and their children concerning their home tasks. We reflected on the experience and grew as a community of mathematical learners.

Game and Puzzle Information:
The following websites contain games and puzzles for you and your child to work on together. Explore the challenges they present that help you and your child apply mathematics.

- Grades Pre-K to 8
    - www.usmint.gov/kids/
- Grades Pre-K to 8 (go to games by grade)
    - www.funbrain.com/
- Grades Pre-K to 5
    - www.mathcats.com/explore/mathcatsgames.html

- Grades 3 to 5
  - http://illuminations.nctm.org/LessonDetail.aspx?id=L576
- Grades 6 to 8
  - http://illuminations.nctm.org/ActivityDetail.aspx?id=40

Additional Resources for Home Use:
To enhance your mathematics home learning environment, the following resources are suggested:

- "Helping Your Child Learn Mathematics" is a free booklet for parents to help strengthen children's mathematical skills and build positive attitudes toward mathematics. Activities are suitable for children in grades Pre-K to 5 and can be found at www.ed.gov/parents/academic/help/math/index.html.
- The website of the National Council of Teachers of Mathematics features *Figure This*, a resource consisting of middle school math challenges for families to explore. This feature also includes a "Family Corner" focusing on parental concerns and questions for all grade levels. This feature can be located at www.figurethis.org/.
- The following books can be found on the Scholastic website at www.scholastic.com/

They are resources filled with activities that are aligned with mathematics curriculum standards, assessment procedures, and instructional practices that nurture the critical thinking involved in game playing and puzzle solving. These books are as follows:

- Math Test Prep That Matters (grades K to 2, 3 to 4, 5 and up)
- Tic-Tac-Math (grades K to 2, 3 to 4, 5 to 12)
- Math Mats and Games (grades K to 2)

Classroom Happenings:
Your collaboration with your child concerning games and puzzles continues to permeate their learning. Activities going on in their mathematics classroom include the following:

- group discussions concerning the interactive homework assignment
- written reflections concerning their collaboration with you

Your Comments:
To open our lines of communication even further, please share your

feedback concerning any part of the parental involvement initiative. Feel free to suggest topics for future initiatives and to express concerns that can be addressed to help you nurture your home learning environment.

Please share your comments below and return to school by_____.

Sincerely,

SUGGESTIONS FOR FURTHER READING

Caldwell, M. C. (1998). Parents, board games, and mathematical learning. *Teaching Children Mathematics*, 4(6), 365–68.

Hinton, J. (2001). *Math works: Mathematical games, puzzles, and diversions for the classroom*. Westbury, NY: Math Matters.

Smith, S., & Backman, C. (Eds.). (1975). *Games and puzzles for elementary and middle school mathematics*. Reston, VA: National Council of Teachers of Mathematics.

REFERENCES

De La Cruz, Y. (1999). Reversing the trend: Latino families in real partnerships with schools. *Teaching Children Mathematics*, 5(5), 296–300.

Guastello, E. F. (2004). A village of learners. *Educational Leadership*, 61(8), 79–83.

Lee, J., Luchini, K., Michael, B., Norris, C., & Soloway, E. (2004). More than just fun and games: Assessing the value of educational video games in the classroom. Presented at CHI, Vienna, Austria.

MacDonald, K. K., & Hannafin, R. D. (2003). Using web-based computer games to meet the demands of today's high stakes testing: A mixed method inquiry. *Journal of Research on Technology in Education*, 35(4), 459–72.

Martine, S. (2005). Games in the middle school. *Mathematics Teaching in the Middle School*, 11(2), 94–95.

National Council of Teachers of Mathematics (NCTM). (2000). *Principles and standards for school mathematics*. Reston, VA: Author.

Shaftel, J., Pass, L., & Schnabel, S. (2005). Math games for adolescents. *Teaching Exceptional Children*, 37(3), 25–30.

Conclusion

I hope the initiatives shared in this book serve you well as you strive to enliven three-way partnerships among yourself, parents, and students concerning mathematical learning. Building parents' understanding of best practices in the mathematics classroom, cultivating parent-child collaboration, and maintaining connections between the mathematics classroom and the home benefit you, your parents, and most importantly your students.

In addition to serving as a resource for parent engagement and home activities, I urge you to consider other uses for this book as well. For example, the mathematics activities shared in chapter 3 are useful in everyday classroom teaching. The games and puzzles shared in chapter 5 are appropriate for a math learning center.

Teacher educators can integrate the format of the initiatives and specific workshops into courses involving mathematics methods and student teaching. The websites shared in chapter 4 would enhance a technology methods course involving websites and mathematics.

In conclusion, I wish to advocate two action steps that should coexist with the initiatives shared in this book. The first being the act of surveying your parents, and the second being the act of reflecting on ways to integrate parental involvement into your everyday mathematics lessons.

The structure and content of the initiatives shared in this book stemmed in part from the voices of parents that were heard through administered surveys. You can learn much from your parents. Just ask. But ask with certain criteria in mind so that you can focus your efforts appropriately.

I share some guidelines here for crafting a survey that would investigate parental needs and perspectives that I have used in a study that targeted areas warranting attention such as the need for more oral communication between parents and children about mathematical thinking (Mistretta, 2007). I encourage you to reflect on these guidelines as you craft your own survey and invite you to review the survey included in the previously cited study.

- Investigate how parents help their child in mathematics.
  - Do they discuss what is going on in mathematics class with their child?
  - Do they help their children prepare for mathematics tests?
  - Do they check to see if mathematics homework is complete and correct?
  - Do they help their child correct mistakes on mathematics homework and tests?
  - What type of setting do they provide for their child to do mathematics homework?
- Investigate how parents communicate with their child about mathematics.
  - Do they ask their children to explain how they arrive at their mathematics solutions?
  - Do they talk with their child about multiple ways to solve mathematics problems?
  - Do they share ideas with their child about mathematics homework/projects?
  - Do they point out to their child real life applications of mathematics?
- Investigate how parents communicate with teachers about mathematics.
  - Do they ask about their child's progress in mathematics?
  - Do they talk about the challenges they face when trying to help their child with mathematics?
  - Do they point out their concerns about the way mathematics is taught?

A common challenge for mathematics teachers is an exhaustive curriculum to complete in a limited amount of time (Mistretta, 2005) and is the reason for my second suggested action step. I urge you to reflect on ways to integrate your parent populations into your everyday mathematics lesson plans.

Doing so can spark ideas for delivering meaningful mathematics experiences to your students in a manner that both uses time efficiently and engages your parents in their children's mathematical learning throughout the year, not just during the initiatives discussed in this book.

I offer at this point a description of just such a lesson that uses time efficiently to connect multiple mathematics standards and engage parents. The lesson concerns three-dimensional geometric solids and the existing relationships among their faces, edges, and vertices.

Students create these solids, namely, prisms and pyramids, and then identify the specific number of faces, edges, and vertices that each contains. They analyze gathered data to determine existing patterns and, ultimately, Euler's Formula, which states the sum of the number of faces and vertices in each of the prisms and pyramids subtracted by two is equal to the number of edges.

The solids that the students create are as follows: the triangular prism, square prism, pentagonal prism, hexagonal prism, septagonal prism, octagonal prism, triangular pyramid, square pyramid, pentagonal pyramid, hexagonal pyramid, septagonal pyramid, and octagonal pyramid. Straws are used to create the edges or sides of the geometric solids, while marshmallows depict the vertices, or corners. The flat surfaces that form are the faces that students can poke their fingers through.

Given such a concrete experience allows students to conceptually understand the geometric terms prism, pyramid, edge, vertex, and face, as well as engage in data collection and analysis. While determining relationships among the numbers, computational skills and algebraic thinking are used when observing the existing patterns among the numbers and the reasons for such relationships.

Connecting mathematics standards in this way is an efficient use of time, but the time spent can become even more meaningful when parents are involved. For example, instead of having the students create the geometric solids in the mathematics classroom, why not have the students create them at home with a family member. Assign each student a geometric solid, and send them home with a set of directions on how to create the solid.

For example, a student assigned the pentagonal prism would be given a set of directions that instructed the student and family member to create two five-sided shapes with straws and marshmallows. They would need to connect the two five-sided shapes with straws so that one five-sided shape was on top of the other five-sided shape causing the base and top of the geometric solid to be the same.

Another example would be a student assigned the pentagonal pyramid. They need to create with their family member one five-sided shape using straws and marshmallows. They would then place a straw in each marshmallow of the five-sided shape so that these straws could be joined together at one common point by another marshmallow. This geometric solid would have a pentagonal base with a pointed top.

Having the students work with a family member in this way connects the home with the mathematics classroom and also saves you instructional time in school. The students can then bring their geometric solids to mathematics class to analyze. There will be more than one of each geometric solid coming into class that lends itself to easy access by the students as they begin to collect their data concerning the number of edges, vertices, and faces contained in each of the geometric solids.

Have the students organize their data into a table to bring home and discuss with their family member. They should look together for patterns and relationships existing among the numbers. Students can then come into their next mathematics lesson with both their observations and their family member's observations that they can share and discuss with their class.

If you didn't involve parents with this lesson, it would take approximately three days. Integrating parents into the lesson causes it to take two days of instructional time, engages parents in the learning process, and allows for both student and parent observations concerning the collected data to be discussed in class. As you can see, one can do much with little time if we reflect first on how to connect standards and involve parents.

Communication is a key ingredient to productive collaboration among teachers, students, and parents. If you find yourself in a setting where language is a barrier, it is helpful to partner parents in a way so that those who speak English can help those who do not.

I have also found it helpful to empower students with the task of communicating what I am saying to their parents. On occasion, translators have accompanied me during workshops and have also translated written documents so that I could distribute materials to parents in their native language. These translators often were parents who offered their services.

Fear of the unknown, as we all know, can cause uneasy feelings. Disarm such feelings by viewing your students' parents as part of your classroom. Engage them in their child's mathematical learning, and use them as resources, not just homework checkers. When you do, meaningful mathematics experiences begin to grow within a partnered community of learners.

REFERENCES

Mistretta, R. M. (2005). Mathematics instructional design: Observations from the field. *Teacher Educator*, 41(1), 16–33.

Mistretta, R. M. (2007). Cultivating parent-child collaboration concerning mathematical learning: A necessary objective for teacher preparation programs. *Teacher Education Yearbook XVI: Imagining a Renaissance in Teacher Education*. Lanham, MD: Rowman & Littlefield Education.

About the Author

Regina M. Mistretta, associate professor in the School of Education at St. John's University, strives to help forward the field of mathematics education through teaching, research, and service efforts that address the needs of students, teachers, administrators, and parents. She teaches both undergraduate and graduate courses as well as directs professional development and parental involvement initiatives concerning mathematical learning.